More acclaim for *The Ptarmigan's Dilemma*:

"Superb. . . . vivid. . . . This book can be savored for both its passionate accounts of the natural world and the informative discussion of the principles that underlie life's order and regulation." – *Library Journal*

"Exceptional. . . . If only more biologists would take the time to write about the 'life and times' . . . in such a profound way." – *Canadian Field-Naturalist*

"Gives a sense of how much joy people can get from going out and looking at nature. . . . The Theberges . . . have clearly led lives of great richness and it is a pleasure and privilege to spend time with them." – *Globe and Mail*

"An important and provocative addition to any natural-history library." – Straight.com

"Have you ever wondered . . . about how life came to be, how it evolved, and how it all works together to give us the astounding natural worlds in which we live? Then this is the book to read. . . . Every page of the book echoes with [the Theberges'] passion and conviction." – *Spirit of the North Magazine*

"If I were asked by a visitor from outer space for the best information on the history and ecology of life on Earth, I'd offer this book." – *Alternatives Journal*

"Anyone who thinks a scientific understanding of nature takes the beauty and mystery out of life, should read this book! And here's a hint – the ptarmigan's dilemma is also yours and mine, because misjudgment about living within our environmental limits can mean death for the bird, and us." – Monte Hummel, President Emeritus, WWF–Canada

"I've travelled to many of the habitats the Theberges write about, but I had no idea what I was missing until I read *The Ptarmigan's Dilemma*. It showed me the intricacy of life as I've never seen it before. Darwin would love this book!" – Bob McDonald, host of *Quirks & Quarks*

"Readable, personal and enthralling . . . essential for non-scientists, but likely of interest also to many scientists." – Eva Durance, naturalist and author of *Cultivating the Wild*

THE PTARMIGAN'S DILEMMA

An Ecological Exploration into

THE MYSTERIES

of

LIFE

JOHN THEBERGE *and* MARY THEBERGE

McClelland & Stewart

Library and Archives Canada Cataloguing in Publication

Theberge, John B., 1940–
The ptarmigan's dilemma : an ecological exploration into the mysteries of life /
John Theberge and Mary Theberge.

ISBN 978-0-7710-8518-5

1. Evolution (Biology). 2. Ecology.
3. Life (Biology). 1. Theberge, Mary T. II. Title.

QH366.2.T44 2011   576.8   C2010-907117-4

We acknowledge the financial support of the Government of Canada through the Book Publishing Industry Development Program and that of the Government of Ontario through the Ontario Media Development Corporation's Ontario Book Initiative. We further acknowledge the support of the Canada Council for the Arts and the Ontario Arts Council for our publishing program.

Library of Congress Control Number: 2010943158

Illustrations by Mary Theberge

Typeset in Caslon Book by M&S, Toronto
Printed and bound in Canada

ANCIENT FOREST
FRIENDLY

This book is printed on processed chlorine-free paper that is 100% recycled, ancient-forest friendly (40% post-consumer waste).

McClelland & Stewart Ltd.
75 Sherbourne Street
Toronto, Ontario
M5A 2P9
www.mcclelland.com

1   2   3   4   5   15   14   13   12   11

For Jenny and Michelle, who have travelled with us on many
of these journeys of exploration

CONTENTS

A TWIG IN AN EDDY

*The trail winds steeply upslope from valley timber to sunlit alpine meadow.
On all sides, the mountains march off into the distance. A marmot whistles, and a mule deer bounds away. There is a wholeness to ecosystems, an
order, an interlocking of species, an unaccountable richness, an intricate
design – without planner, engineer, or architect.*

But there is an operating manual.

IN THE SOUTHWEST YUKON stands Amphitheatre Mountain,
a low mountain in comparison with the looming giants to the
west. Early in our careers together as wildlife biologists and naturalists, we came to know this mountain well. We often camped
there, in summer sunshine and fall snows, because it provided a
tundra vantage point to observe woodland caribou whose ecology
we were trying to understand. The caribou moved in and out of
nearby Kluane National Park, and in those years their numbers
had dwindled.

Amphitheatre Mountain features a flat, plateau-like top that can be reached by a stiff climb up its western slope. White Dall's sheep graze its lower flanks, keeping its eastern crags in view as escape terrain, and peregrine falcons use those crags for eyries.

But most significant to our story, the mountain features the remains of a stone forest. One day we watched a golden eagle swoop down and land on a rock protruding from the ground. A rock? It looked more like a stump, but this was treeless tundra. We investigated and found it *was* a stump – petrified. Only then did we realize that many of the rock outcrops around us were actually petrified wood, including the one we'd used as our favourite perch. Some of the wood had been transformed into coal; some was an off-white that clearly showed ancient tree rings stained brown.

The stump had been a metasequoia, remnant of a warm, wet forest that grew in the Yukon some 35 million years ago when the land was an inland plain, before the coastal ranges rose. Planet Earth was once believed to be completely stable, but now we know that, on its tumultuous journey through time, its amoeboid surface contorted as mountains rose and eroded away, sea floors appeared and vanished, continents split and drifted apart, rivers were born, carved canyons, and disappeared. And through all its transformations has streamed a panoply of life.

Those fossils, and others we have admired in situ – pumice rocks from Washington State etched with 48-million-year-old maple leaves; pieces of Wyoming sandstone with 50-million-year-old fish imbedded in them; chunks of dark red 150-million-year-old petrified wood from Arizona; a coprolite (petrified dung) from the intestines of some unknown coyote-sized early mammal or dinosaur – all whisper about a time before humans were around to appreciate, or wonder, or piece together how the self-organization and persistence of life ever came to be. How has life kaleidoscoped through the ages to arrive at what we have today? How, with no endangered species act, no conservation lobby to express outrage at environmental calamities, no environmental policies, did living things struggle through adversity and endow us today with a rich biological inheritance unparalleled in the history of life – "without planner, engineer, or architect"?

This book is a journey of discovery of the ways life adapts, persists, and is able to organize itself at all the different levels of existence, from genes to whole ecosystems, as seen through the eyes and experiences of wildlife ecologists. At each of the levels of life's organization, novel mechanisms have accompanied new opportunities as life, behaving like a twig in an eddy, has seemingly poked and probed until it found a way to continue its journey. What are the mechanisms? How do they work? And why?

Any good journey leads somewhere, and we found ourselves arriving at several destinations. One is the satisfaction of a holistic view of life, in which some of its mystery has been supplanted by an appreciation of how things work. Another destination is a platform for viewing, participating in, and helping to resolve the plethora of environmental issues that face the world today. The problems are severe, from disappearing species and the consequent impoverishment of life to the instability caused by climate change, pollution by toxic chemicals, and the one force that increasingly dictates a different future for life in all forms – us. Other books have delved into the details of particular environmental problems and their possible solutions. Our book attempts to answer the more fundamental, underlying questions and provide a science-based view and comprehension of life. The final destination of our journey is a perspective, a prognosis, for the future of life on Earth.

Our early married life was a time of heightened environmental concern driven by the shocking scenarios described by such writers as Rachel Carson in *Silent Spring* (1962) and Paul Ehrlich in *The Population Bomb* (1968). Everyone wanted to be an ecologist in the 1960s. But there was a hole in the environmentalism of the day, an unsettling lack of understanding about the ways life is organized, an absence of any unifying or central concept of how life supports itself. We were like surgeons operating on a heart attack victim without understanding the basic circulatory system. While money poured in for environmental impact assessments, little was made available for basic research to plug that information hole. We, like some of our colleagues, had some disturbing experiences with impact assessments, naively and inadvertently covering for developers who really didn't give a damn about the environment.

But the impact assessments we engaged in, along with the pure research dollars we managed to scrounge, allowed us to experience some superb wild country. So our lives have been sprinkled with experiences that have repeatedly raised the same basic questions: How is life's marvellous self-organization accomplished? When and why might it fail? What of the future?

Today, scientific knowledge in many fields has accumulated to the point that these questions can be addressed more effectively than was possible in the early days of environmentalism of the 1960s. These are exciting times. Unparalleled advances are changing how we understand the living world. New discoveries are uncovering the ways living things are hooked together (ecology), and the processes that got them there (evolution). Firing them has been an array of technological advances, from computers that allow us to squeeze added meaning out of data, and telemetry that reveals the precise movements of animals, to satellites that map ecosystems from afar.

These are also crucial times. Never before has a population of any species tried to rejig so much of the world to its own ends. The result is unprecedented pressure on the global systems that support life. We need to detect the land mines.

Yet, these are deeply satisfying times, too, when an appreciation of the beauty and mystery of the natural world is increasingly grounded in science. The forefronts of science keep coming at us like incoming waves. The natural world operates with both tricks and surprises, force and fury. It is both astonishingly complex and simple.

Some discoveries are novel, like the "order for free" hypothesis that challenges the sole supremacy of natural selection. Others are unexpected, like a previously unknown method of solar energy capture. Still others are refurbished, such as the unexpected nuances of the formerly discredited theory of the inheritance of acquired characteristics. And some warn us of a destabilizing biosphere, like the diatom die-off in parts of the Southern Ocean.

From this rich mix of ideas and concepts have arisen new ways of interpreting the operating manual of life, new ways of perceiving the biosphere – that thin, living cellophane wrapped around the globe, no thicker relative to the Earth than the skin is to an apple.

On our honeymoon canoe trip in Ontario's Algonquin Park, we awoke one morning to the howls of wolves drifting to us through the mist from the far end of the lake. Hastily crawling out of the tent, we launched our canoe and paddled through the fog to a vertical rock bluff that rose from the water. There we idled in the silence of early morning. A distant loon called, and we heard the weak peeps of a few migrating warblers. Gradually, sunlight began working through the mist, and as it opened on the top of the rock ledge, out stepped three wolves, close enough that they could have easily jumped down into our canoe. Initially they exhibited no fear, maybe because we were still partially shrouded. Then they jumped back, and the ethereal moment was gone.

They raised a question as haunting as their surroundings. How do you perceive a wolf in a natural environment? Is it an actor on a stage? Or is it part of the stage itself? We still ask those questions on each research trip to Yellowstone National Park, where we study wolves and watch their life-and-death dramas played out across the sagebrush hills. The answers change the way you perceive the world, putting the emphasis on what is important in a different place, making you read nature differently.

The questions arose again and again in our fieldwork. The beauty of the wood ducks, which animated the pond each spring at the back of the woods where we lived, begged for an explanation. And the pulsating life in the Indonesian jungle, whose interactions embraced everything from soil nematodes to flying foxes, fairly shouted the questions: How did such a living machine ever come to be? And what has kept it working? And wildebeests stretched across the Kenyan savannah in their hundreds of thousands, and the similar superabundance of sandhill cranes in the sky and on the sandflats of Nebraska's Platte River – each exposure to such spectacles has raised the same nagging question: Is there a central theory of life support, an overarching answer to the seemingly incomprehensible self-organization of nature?

We have looked for answers primarily where we conducted field studies: in the verdant hardwood forests of southern Ontario and the dark boreal forests farther north; the species-rich mountains and plateaus of interior British Columbia; the rolling tundra of central Alaska; the dry plateau country of Washington, Oregon, and California; the sprawling taiga of central Labrador. We have looked, too, in far-away lands: Africa, Indonesia, Australia, New Zealand, Mexico, Central America, Antarctica. Sometimes we have investigated with the help of research grants and of colleagues and graduate students. Other times we have simply visited as curious naturalists.

The questions we have asked can be pursued only by weaving the twin subjects of ecology and evolution together. To a large extent, the two are simply different time-perspectives on the same thing. Ecology attempts to explain what is happening now in nature. The woodlot down the street is not just "there." It is wildly active. Organisms are playing out their lives in dramatic interactions. Energy is flowing, nutrients are cycling, and species are struggling, co-operating, tolerating, adapting, and exploiting one other. The amazing product of all this activity is a functioning ecosystem.

Evolution explains what happened before, leading up to the present state. The word means no more than "sequential change," which was well recognized to occur in nature long before Charles Darwin. Yet still today evolution is poorly understood and even rejected by a large proportion of society. A recent poll revealed that half of Americans deny evolution,[1] and the proportion of Canadians is likely similar. Some people believe that humans were created about 6,000 years ago. Others may not have thought much about the roots of humanity at all. Both sets of people must tune out all news of science, because these poll results duplicate those found two decades earlier, despite major scientific advances, for example, in genetic engineering that allows manipulation of the very stuff of evolution, and in radioisotope dating that provides an overwhelming confirmation of the antiquity of life.

If you are among those who have not started up the staircase of science, you could join the exploration here. And exploration it is, because while evolution is indisputable, even a logical necessity, the mechanisms that drive it are still open to new insights. With each discovery, the questions just keep getting better. There is still overwhelming mystery.

Many students of wild places and things can recount singular events that not only helped orient their careers but also became benchmarks for their interpretation of later experiences. Our benchmarks invariably influenced the interpretations we set out in this book.

A defining event for me (Mary) came during a sabbatical at the Grand Canyon. Somehow, "grand" does not successfully describe this amazing creation. This is not one canyon but a composite of thousands. Scenes continuously shift as light plays off the rocks, creating shadows and contrasts. Sunsets and storms lend added dimensions. The permutations are unceasing, the moods without end. Here is a place that humbles the soul, that makes mincemeat of human arrogance. Here is a record of geological history laid open for all to see, bearing mute witness to violent events of mountain building and lava flows, of ancient seas, water, erosion – and time. Eons and eons of time.

We spent many days down in the canyon, but our first trip was the most memorable. Slowly, we picked our way across the rocky talus slopes on our descent into the depths, hoping that a misstep would not hasten our arrival at the bottom. As we dropped down the primitive Hermit Trail, on display and parading the progression of time were the horizontal layers of sedimentary rock that the Colorado River had sliced like a band saw. That evening, at our isolated campsite on the basement rock, we pondered the ancient landscape, when the Earth was only half as old, when there was no canyon, no river, and of course no terrestrial life.

As night settled over the Hermit Basin, a full moon outlined the cliff edges and floodlit the western walls. The distant roar of the river blocked out all other sound. Lying in that moonlit hollow, where everything visible was inanimate, life seemed no more than a thin veneer, a shadow, an afterthought. In a slow, timeless way, the moon shifted its illumination, as it always has, and will whether we are here to witness it or not. When daylight poured down from the east, it was decidedly reassuring to see the desert ecosystem once more, rich in wildflowers and shrubs, and the tracks of antelope squirrels and mule deer in the sand. Yet, the impression remained that life clings precariously on planet Earth with little leeway between its richness and abundance, and extinction. It was a profound message of both pricelessness and fragility.

A benchmark for me (John) came one spring evening in my final undergraduate year at the University of Guelph, Ontario, when I went to hear a lecture by botanist Dr. Hugh Dale. The lecture was billed as an exploration of the relationship between science and an attitude towards life – not a novel topic, but every young mind must grapple with it anew. Dr. Dale's lecture that evening touched only lightly on Darwin's and Wallace's great discovery of natural selection and the debate and furor about it that persists even today. Instead, the part I have never forgotten through the years was his blackboard illustration of a staircase. The staircase represented science, leading nowhere, just ending, with no landing. Up it gropes the student of life, feeling his way on hands and knees in the dark, like the hero in Robert Louis Stevenson's novel *Kidnapped*, climbing perilously along an outside wall, the scene illuminated periodically by flashes of lightning. The steps are uneven. Some represent "quantum leaps," such as the discovery that the world is round, or that the sun is at the centre of the solar system, or that evolution is occurring, or that the continents are drifting around. Other big steps in the staircase represent science-based theories concerning the origin of the universe, the birth and death of stars, the origin of life.

At the top of the staircase, Hugh Dale drew some wavy, upward-sloping lines. That is where the student crouches in the darkness, reaching with his hands into space, probing for the way forward. Beyond science is speculation, now and forever, because all science can do is hammer out some more steps. The best scientists stand on that top step of knowledge and know there is much more. The best view of life, one's personal philosophy, comes from standing there, too. The highest step, for us, for many people, is attempting to understand the organization of the life support systems of the world we live in.

This book is meant to contribute to that highest step through pursuit of its central questions: How is the self-organization of life accomplished? Is there a unifying concept involved? And what do the answers say about the future of life on Earth? It is our hope that care, concern, and compassion for the natural world can be spawned *with a special conviction* when they are based on knowing as much as possible about its scientific underpinnings – not just scientific facts, which are only the starting place, but science as a rational springboard for emotional connection.

The book starts with a close-up on life in its essential, universal form – molecules – and builds from there, as life does, to complex structures – individuals, populations, whole ecosystems – and finally to global life support systems and what they tell us about the future.

SPECIES ADAPTING, ADJUSTING

IN PART I, in which we examine the lowest level of biological organization, we sift through our wildlife experiences with various species – wood ducks, songbirds, elk, moose, and wolves – to grapple with the question: What are the underlying mechanisms that organize raw molecules into species, together with their staggeringly complex and finely tuned adaptations?

For answers, we look in several places: conventional explanations of evolution; recent novel extensions of it; examples of biochemical and behavioural underpinnings; and failures or compromises. These examinations both provide answers and lay the foundation for our ongoing exploration up the biological hierarchy towards an integrated understanding of the self-organization of life.

CHAPTER ONE

WHERE THE WOOD DUCK GOT ITS BEAUTY

Years ago, the writer and anthropologist Loren Eisley sat with a friend on the bank of a pond watching a wood duck. Turning to Eisley, the friend asked, "Do you really think the wood duck got this beautiful plumage by blind chance?" Out of a chaos of swirling stardust – atoms, electrons, protons, neutrons – can one fathom the creation of a wood duck without a creator? Eisley had no answer – but there is one now.

THE WOOD DUCK KNIFED through the bare limbs of the hard-wood forest at a reckless speed and hauled up, wings cupped in a vertical stall, on a thick branch of a dead elm. Behind it meteored its mate, twisting this way and that before landing beside the drake. For a minute, two minutes, they perched there, parallel to the limb, very un-duck-like. Then they edged their way out farther and suddenly dove off through the aerial tangle and out of sight.

That they did not end up impaled by a branch or splattered on a tree trunk is one of the marvels of wood duck skill.

But there is another marvel about wood ducks, one that drives at the very heart of life and living things. Not that the marvel is unique to this species, just extreme, which makes the wood duck a symbol, a standard-bearer. Maybe other ducks, maybe all nature looks in awe at the wood duck, envious of its achievement, for the wood duck represents an example of the most, or the best, in gaudy beauty that nature can achieve.

For twenty-five years, every spring we hiked to the silver maple swamp at the back of our Ontario woods to ponder anew the imponderable, to work through the evidence once more. The spring woods is a good place for that, with life bursting anew, with hepaticas and trilliums poking up and the intricate songs of ruby-crowned kinglets and winter wrens priming the breeze. But the trigger for this annual event was the wood ducks.

They would arrive before the leaves had budded, while the silver maples were decked out in tiny, crimson flowers. We would glimpse them after cautiously stalking along the edge of the swamp, our footsteps muffled by last year's sodden leaves. They would be there, swimming alertly among the logs and stumps in the helter-skelter fashion that is uniquely theirs. The dark water down in the swamp would reflect their mirror image.

The male wood duck is the most brightly coloured of North American waterfowl, with red and white on its beak, a green cape over its head, cinnamon breast, and yellow flanks – a floating palette of colour. In the first Canadian bird book, Alexander Milton Ross described the "summer duck, or wood duck" as "without exception the most beautiful of all our ducks."[1] To be the most beautiful is an accomplishment because of other elegant challengers, such as the hooded merganser, with its white crest edged in black, its dark back, and russet flanks, or the canvasback, with its cinnamon head and red eye. But the wood duck, because it is overdressed even for a carnival, is the species that fairly shouts out, "How did I become so elegant, so richly coloured, so beautiful?"

The question nagged at us. Most people are curious about life's complexity and beauty. Observing any being can prompt questions about how its attributes came to be. Take the veery, for example, a small, brown, nondescript thrush, the antithesis of splendid colouration. But have you ever heard its song, or a choir of them singing at dusk in the deep shadows of the forest? Many evenings in various Algonquin Park white pine cathedrals, while sitting beside a campfire or lying in the tent, we have listened to the veery ensemble, best of all choirs. We have recorded their songs, and when played back at half or quarter speed their liquid beauty and complexity is astounding. Some might say that if that song came about by chance, then so must have Beethoven's violin concerto.

Then there is the Polyphemus moth, an insect of elegance, displaying tawny wings banded with grey, white, and touches of red, and sporting a single large "eye" on each hind-wing. The "eye" is a perfect mimic of a large vertebrate eye, complete with a fake pupil, but constructed out of wing scales instead of the stuff of a functional eye. What about the splendour of a pink, deep-throated calypso orchid, intricately designed for insect pollination? Or the vertebrate eye, a marvel of engineering? Or the human brain? It's little wonder that people recoil at the thought that living things, including us, have come about by chance.

Yet, hold on. "Chance" is a loaded word, as any Mississippi gambler knows. Could it be that the deck of life is stacked, the dice loaded, the game rigged? Is it possible that rules and regulations exist, to be adhered to or broken? Might winners be rewarded, transgressors cut down? Maybe a wood duck's bright plumage was "in the cards" all along.

Natural selection exerts the most immediate and severe curb on chance, the picking and choosing by the environment of the individuals that are best fit, or, more accurately, the weeding out of the less fit. It operates with just two requirements: replication and variation. The former is met by the production of offspring, the latter by their physical differences. Those individuals whose differences confer some adaptive advantage, be it environmental fitness, competitiveness, hardiness, or longevity, will invariably leave more offspring. Populations will become dominated by them. Evolution – sequential change – will have occurred. It is as intuitive and simple as that. It has been verified experimentally in many ways, such as in a long history of crop and livestock improvement.

How natural selection has led to the wood duck's plumage, however, involves a special twist. In spring, wood ducks spread out across the southeastern, central, and western sectors of North America, where they occupy small ponds and wooded swamps. At that time, wood duck drakes are decked out in full splendour, as are all ducks. After breeding they enter a post-nuptial moult, becoming drab brown and staying that way through late summer and early fall. So do most songbirds. However, a telling difference between ducks and songbirds happens next. In late fall, ducks moult back into their bright plumage, while most songbirds wait until spring. Predictably, display and pair bonding begins in both groups as soon as they are dressed appropriately for the occasion. This difference in timing illustrates that the selection pressure that drives bright plumage is mating, with its dual requirements of attracting a female and out-competing other males.

In male vertebrates, a great variety of anatomical features help play this role: crests; plumes; fanned tails; brightly coloured eyes, beaks, legs, and feet; horns and antlers; and large body size – all developed to attract female attention. So pervasive and obvious is selection for mating purposes that Charles Darwin termed it "sexual selection," differentiating it from natural selection, which he related more closely to survival. He defined sexual selection as "the advantage which certain individuals have over others of the same sex and species solely in respect to reproduction."[2] Sexual selection, then, a special type of natural selection, has been the dominant force driving the plumage characteristics of wood ducks.

The significance of sexual selection in creating display features in animals does not mean that natural selection is less functional. Heavy persecution has illustrated natural selection at work, too. For instance, the selective effects of severe trophy hunting has caused shrinkage in both horn and body sizes in several species, such as bighorn sheep in Alberta.[3] Wood ducks also have been heavily hunted, so much so that further shooting was banned between 1918 and 1941 and limited for years after that. However, unlike mammals with horns, hunters do not select which wood duck to blast on the basis of its appearance. If they did, then selection would have taken place and wood ducks would have lost some of their beauty, because the price of being more beautiful than other wood ducks would have been an earlier death with fewer or no offspring.

Conservation measures, primarily limitations on hunting, have brought wood ducks back. In my (John) student years they were still recovering, and part of my job for the Metro Toronto Conservation Authority was to survey their breeding success at nesting boxes provided to owners of private ponds. After a lot of ladder climbing and more than a few falls, it turned out that most of the boxes were unoccupied. Then, one autumn evening twenty years later, as we traversed a dike out in the broad Long Point marsh on the north shore of Lake Erie, wood ducks seemed to fill the sky. They were on migration, congregating at the mouth of this land-funnel that extends out into the lake. We watched flock after restless flock skitter across the orange sky and dip into the marsh for the night.

Today, the estimated fall population of wood ducks in North America is between 2 and 4 million after the hunting season, which still slices out more than 1 million and is the principal cause of death.[4] However, the wood duck population seems to be reasonably secure.

Thus, natural and sexual selection provide the immediate answer to the question of where the wood duck got its beauty. Selection has curtailed chance by drawing only certain cards from the deck that suit its purpose – to leave more offspring. Today, as Darwinian natural selection receives ongoing and even increasing attention from biologists, especially with the advent of new techniques in genetics, its basic tenets, and power, are repeatedly reconfirmed.

Chance is more closely associated with genes than with natural selection. Genetic variety comes first, then nature selects from it. But where did the genes originate that dressed the wood duck that way? Again, the deck is stacked, but with greater cunning, a more subtle sleight-of-hand.

Here we enter the laboratory world of the geneticist working, ironically, with both the very basis of life and at the same time only its abstraction. To correct for it, we gave a photograph of a wolf to our geneticist partners to hang on the wall so they could look up from their microscopes now and then and remind themselves of what they were studying.

Geneticists have made remarkable gains in explaining how genes work. But the field is relatively new, only some hundred years old, and misinterpretations happen. The pioneering work of Gregor Mendel with variously coloured peas led to the idea of simple inheritance – one gene equals one biological trait. However, through mapping of the human genome and similar projects, it turns out that most genes are linked, reducing chance, as if some cards in the genetic deck were stuck together. Other genes simultaneously control several features of an organism, or form complexes, or play regulatory roles by turning certain genes on or off. Geneticists today puzzle, for example, over phenomena such as "hox genes," which are clusters of genes that express themselves as an overarching group during the development of an organism.[5]

Interacting genes raise the possibility that each characteristic of a wood duck – its red eye, its green helmet – did not come about independently. Interacting genes could have acted to cause the wood duck to suit up in its particular way, without hypothesizing the chance appearance of many individual and unrelated genetic events, just as we do not buy our clothes thread by thread, but as a jacket, blouse, or pants.

The seamy biochemical world of DNA is a complex environment of interacting atoms and molecules that respond to the dictates of electrons, protons, neutrons, and sub-atomic particles. Some interactions allow chemical bonding, others result in rejection. But despite this complexity, two primary patterns emerge: recombination of existing genes, and mutation that forms new genes. Recombination, the more important of the two in providing genetic variation,[6] is the shuffling of existing genes from generation to generation. Chance dictates these rearrangements but it is curbed by the limitation of what is there. By analogy, there are only fifty-two cards in the deck, not some larger number, and each card has a specific face value. These impose limits. However, in the deck of genes, now and then a chance mutation will show up, a new card shuffled in.

The genetic deck would be stacked even more if the French biologist Jean-Baptiste Lamarck, who advanced the infamous theory of the "inheritance of acquired characteristics," had been right. For decades he has been set up in the history of science as Darwin's stooge. However, he actually foresaw evolution, which in his day was called "transmutation," meaning one form changing into another.[7] Lamarck argued against the established theory of the immutability of species. In a way, he provided a starting point for Darwin, who then had only to get the central mechanism – natural selection – right.

Lamarck's mechanism, the "inheritance of acquired characteristics," has long been decried as a mistake. An example given in both old and new genetics textbooks is of a blacksmith who builds up big biceps from swinging a hammer. He will not give birth to children with big biceps, as Lamarck supposed. The father's occupation cannot influence the genes he passes to his offspring. A giraffe that stretches its neck to reach higher up in acacia trees will not produce young with longer necks as a result. Too bad. Many of the marvellous adaptations of living things would be considerably more plausible if such a direct feedback from an environmental need to an anatomical structure were true.

But wait. Lamarck's theory has life in it yet. In 2000 an article appeared in the prestigious journal *Science* with the title "Was

Lamarck Just a Little Bit Right?"[8] The article detailed ongoing research that explains how the environment can call various genes into action, or silence them, with down-the-gene-line inheritable effects. (This new field of discovery is explored in Chapter 2.) However, at this point in the story, it is enough to recognize that genetic chance is curbed by most genes hanging out in gangs and acting in coordinated ways, and by the deep physics of the lock-step molecular waltz.

As well as natural selection and genetics, helping to explain the wonder of the wood duck is *time*, another way that chance is reined in. For change to happen and eventually build intricate biological things – a wood duck's plumage, a veery's song, a vertebrate eye, the complex physiology of our own bodies – takes time, deep time, immense caverns of time. Darwin realized that and went to his grave puzzled by how evolution could have proceeded from unicellular organisms to complex animals in just 100 million years, that being the accepted age of the planet in his day.

The time involved in biological evolution is unfathomable. With enough time, with each reshuffling of the genetic deck generation after generation, the chances of a seemingly impossible anatomical structure emerging increase. Impossible may become improbable, then only unlikely, then, reshuffled over and over again, may become possible, even probable – and then the structure appears. Play till you win. But the process of winning through natural selection is an incremental thing, like a gambler amassing his earnings gradually, hand after hand. Each slight improvement that is advantageous is retained and built upon. This key point in understanding evolution through natural selection explains why the improbable construction of wonderful things, like the vertebrate eye, or the wood duck's plumage, is not an un-fathomable engineering feat. Each was, instead, the consequence of tiny, incremental improvements over long reaches of time.[9]

Glaciers left most of the northern hemisphere 10,000 years ago, the first human showed up 2 million years ago (depending on where you draw the Australopithecus-Hominid boundary), the final rise of the Rocky Mountains happened 5 million years ago, dinosaurs died out 65 million years ago, the first deciduous trees appeared 120 million years ago, life left the seas and crawled up onto land 440 million years ago, the first life on Earth emerged between 3.5 and 4 billion years ago, our planet formed 4.5 billion years ago, the universe originated 13 to 15 billion years ago. Few people have these dates memorized, maybe because we cannot comprehend such vast stretches of time. In the vacuum caused by this lack of comprehension lies the difficulty in accepting that, with enough time, remarkable things are possible.

Time has worked on the lineage of ducks, just as it has on all biological lineages. The average length of time for a new species of bird to emerge has been estimated as between 3.3 and 5.5 million years.[10] Although recognizable change can happen in just a few years, the rapid evolution of any species must be accompanied by several conditions – such as quick, almost complete population turnover and genetic isolation – and is an exception, not the rule. No wonder species look immutable, just like the position of the continents. The world, however, is not how it seems. Things change.

The closest living relative of the wood duck is the mandarin duck of Asia, India, and southern Russia.[11] Male mandarins are brilliantly coloured in ways that show their relationship with wood ducks. They too have white shoulder crescents and yellow flanks, but they have large white patches on their heads. Females of the two species are almost indistinguishable. Wood ducks and mandarin ducks are the only two species in their genus, *Aix,* and obviously had a common ancestor that has been lost in time.

After the mandarin duck, the wood duck's closest living relative, based on admittedly incomplete evidence, appears to be the hooded merganser,[12] with which it shares a few characteristics, such as relatively small body size, a crest, a preference for small ponds and semi-wooded areas, and the habit of nesting in tree cavities. Genes program behaviour, too.

How long ago the split occurred that separated genus *Aix* from an ancestral line is unclear, but a proposed date for an earlier split between ducks and geese/swans based on fossil evidence is 25 million years ago.[13] By 15 million years ago, many genera of modern-day ducks were present. Genus *Aix* likely appeared somewhere at the earlier end of this range of dates. Somewhere back then, the wood duck got its beauty.

Back then, the world was different. It was hotter. Even though many species of mammals were similar to those of today, wood ducks would also have run across four-tusked gomphotheres and short-faced rhinoceroses wallowing in the mud at pond edges on the central plains. Then, in the Pleistocene Epoch, over the past 2 million years, they witnessed sabre-toothed cats stalking mammoths, and gigantic ground sloths reaching high up into trees to browse. Wood ducks persisted despite the comings and goings of glaciers, the kaleidoscope of environmental change, and the appearance of humans. Wood ducks can boast of a robust, time-tested design.

To recap: natural selection, genetic variation, and time are the three necessary ingredients for the evolution of the beauty and complexity of life, a fact we have known for more than half a century. Yet, each spring, as we looked at the year's first wood ducks in the silver maple swamp, we wondered if that is all there is to it. And now it seems that there is another ingredient at play. Possibly, we are on the edge of a new and expanded view of life. Natural selection, genetic variation, and time may have an accomplice. Like Ali Baba, we are just now opening the door and glimpsing a large and beautiful chamber within, one that will take years to explore. The sign on the door reads "Order for free."[14]

A hint of the existence of this other ingredient was provided in 1949 by the American paleontologist George Gaylord Simpson in his seminal book *The Meaning of Evolution*.[15] The first two-thirds of the book are devoted to the overwhelming evidence for evolution. Simpson's great accomplishment was to work out relationships among fossil mammals across the 200 million years of their existence. His even greater accomplishment, however, is found in the last part of the book. In Simpson's day, as today, evolutionary understanding was founded on what is known as the Modern Synthesis that married Darwinian natural selection with genetics. But, having explored how evolution happens, Simpson went on to ask: Where, if anywhere, is it going? In asking this question, he laid the foundation for the fourth ingredient of evolution.

Simpson rejected "finalism," for example, which held that life is progressing towards some utopian state, towards heaven on Earth. Fossil evidence showing the prevalence of extinction and blind alleys refutes that. As well, he rejected an alternative viewpoint, called "vitalism," which postulates the existence of a unique life force, because none has ever been found. Instead, Simpson concluded that what separates life from non-living things is organization alone. Both are comprised of the same elements found on the Periodic Table that adorns the walls of chemistry laboratories the world over. Involved in both, too, are the basic forces or laws of physics. In living things, however, the elements are simply put together differently, in a self-replicating and metabolizing way – self-organized rather than directed by an unseen force.

Today, evidence is mounting to support Simpson's conclusion that spontaneous order or pattern in nature is generated by inter-acting parts operating together to form *systems*, whether those parts be molecules in rock, sand, clouds, a chemical flask, or living things. For example, an automobile engine is a *system* that oper-ates by virtue of interacting parts. So is an individual organism, a population, an eco*system*, a solar *system*.

If order does emerge, fuelled from the sun, where does that order come from? What does it mean for evolution and life? For some years, to introduce students to these questions, I took two objects to my third-year ecology class for what I called the annual "shell game." One was a porphyry olive shell, which is a ten-cen-timetre-long marine member of the snail group. It is coloured off-white with an intricate and beautiful pattern of spiderweb-thin, brown symmetrical scales. We had picked it up on a seashore in New Zealand. The other object was a geode cut in half to reveal its inner crystalline core. That had come from Arizona desert country, a remnant of an ancient volcanic explosion.

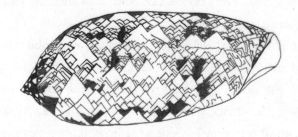

I held up the shell first, with the question: "What is the origin of its beautiful and intricate pattern?" Always someone would answer, "natural selection," an understandable response, because that part of the course dealt with natural selection. Discussion would follow about protective colouration on a shallow sea floor where light dappled in, or similar adaptation to one thing or another. Then, I brought out the geode, and immediately obvious was that natural selection did not apply. It was an inanimate object. But it also displayed order and beauty. Did that come about by chance?

Pattern formation is well known in chemistry and physics. The interacting subunits are inanimate objects such as grains of sand or chemical reactants. Pattern, such as on the surface of boiling water or oil, or the sand ripples on a beach, is created through interactions based solely on physical laws. Pennsylvania biologist Scott Camazine, senior author of a book on self-organization, explains, "The molecules of oil obey physical laws related to surface tension, viscosity, and other forces governing the motion of molecules in a heated fluid. Likewise, when wind blows over a uniform expanse of sand a pattern of regularly spaced ridges is formed through a set of forces attributable to gravity and wind acting on the sand particles."[16] Similarly, the ordered, crystalline structure of a geode is founded on physical laws such as gravity and the forces that operate in atoms that bond silica and oxygen into molecules of silicon dioxide and associated minerals in a certain, structural way.

What about the shell? Biological systems must obey the same physical laws. All matter is so bound. Why can't whatever created order in the geode also have worked on the shell? Instead of attributing the shell's pattern solely to natural selection, more likely natural selection has operated later, on pre-existing order brought about by the nature of complex systems – physical, chemical, or biological.

The proponents of spontaneous order, like Stuart Kauffman and Brian Goodman, have thought a lot about biology, too. When asked where on a scale of one to ten they would rank natural selection in its importance in creating biological order, with ten being highest, Goodman responded, "Close to one."[17] That is revolutionary! Most biologists would score it up around eight or nine, or even ten. The view of people espousing spontaneous order is that natural selection works on pre-existing order. Natural selection, then, explains how organisms fit their environment, but is not sufficient by itself to explain the existence of order.

Darwin himself sensed a basic, pre-existing order. Daniel Dennett, in his book *Darwin's Dangerous Idea*, attributes to Darwin the words: *Give me order and time, and I will give you design. Let me start with regularity – the mere purposeless, mindless, pointless regularity of physics – and I will show you a process that eventually will yield products that exhibit not just regularity but purposive design.*[18] Order, then, is mere pattern. Design is pattern put to a purpose. Natural selection achieves that purpose.

For self-organization to emerge without the benefit of external blueprint, plan, or leader, there must be some rules. Discovering rules has been at the heart of the advances in understanding self-organization, because with a computer, it is reasonably easy (for the computer specialist) to simulate their effect. Start with unorganized dots or blips or any symbol put up on the screen. Establish a set of rules that describe when the symbol will be turned on or off by virtue of what its neighbours are doing. Let the program repeat itself over and over, so that it mimics biological generations, and pattern emerges, to repeat itself indefinitely.

Imagine our surprise and satisfaction one day in finding a picture of a porphyry olive shell in Camazine's book describing self-organization. He discusses the work of a biologist named David Lindsay, who was able to simulate the exact pattern of that shell on a computer by establishing a set of random points and imposing on them just a few simple rules.[19] The rules were simple instructions that he programmed about the frequencies of different combinations of vertical and diagonal lines that occurred in the shell pattern.

Now, envision similar simple structural rules being applied to the DNA-molecular gel in a reproducing cell. Pattern will emerge. Complex rules may not be necessary. As described by Camazine, "The richness of structures observed in nature does not require a comparable richness in the genome but can arise from the repeated application of simple rules by large numbers of subunits."[20]

Where do the rules come from? Much more remains to be discovered, but natural selection plays an obvious part. If rules do not work for a species, natural selection will purge them from the population. If rules work, they will persist. Camazine says, "Natural selection tunes the parameters of living systems to avoid chaos. In most situations, it would probably be grossly maladaptive for a living system to exhibit chaotic, disorganized pattern."[21]

Such a view of evolution, with a part played by inherent order worked on by natural selection, is comforting; any mechanism that reduces total blind chance in evolution is comforting. However, the validity of this concept comes not from the comfort it provides, but from objective analysis and synthesis – in that fundamental way, science departs from dogma. Enough evidence has accumulated to make the idea of spontaneous order in biological systems convincing.

So we return to the wood duck's plumage, an ordered product of greatly whittled-down chance. The marvellous adaptations of living things come about partially from pre-existing order operating in or on genetic systems with rules established, in part, by physical laws and, in part, by natural selection. In addition, natural selection operates on individual traits after they are formed. That dual role for natural selection justifies a higher score than the "one" given by the proponents of spontaneous order for its contribution to order and design, raising it to at least five or six, maybe higher.

Add time, generation-by-generation adjustment, and occasional mutations. Add more time. Bright male plumage is favoured by females in their choice of a mate, which provides positive feedback to go on adjusting incrementally in that direction, possibly with the help of some linked or interacting genes creeping in. And behold, a drake wood duck! Based on what we know today, that, roughly, is how it got its beauty.

Undoubtedly, there is more to it. If ours is a complete explanation, then science is at an end. Instead, in every field, science progresses. One discovery leads to another, the questions becoming more fundamental all the time. That is what makes science exciting.

We eventually sold our home in southern Ontario and moved to British Columbia, so we can no longer walk to the silver maple swamp and ponder the secrets of wood ducks. We still visit our cabin near Algonquin Park, however, nestled beside a beaver pond studded with standing and dead trees – just the place for wood ducks. They nest there, and each spring morning above the raucous chorus of spring peepers and gray tree frogs we can discern their high, squeal-like calls. We used to canoe out to them for a better view, but our pond, like all beaver ponds with enough time, has partially drained. Still, there is sufficient water for at least one pair of wood ducks. When we return next spring for our annual visit, we will be looking for them.

CHAPTER TWO

BIG ELK, LITTLE ELK

*Neither providence nor history has been kind to Jean-Baptiste Lamarck.
Despite his 1809 breakthrough theory of transmutation, which shook the
biological world, he died penniless and blind and was buried in a pauper's
grave. But now a question that has bedevilled ecologists – namely "What
is a species?" – has generated exciting new discoveries and helped resurrect his reputation.*

ELK IS THE NAME North Americans give to a long-legged,
curved-necked, ungainly creature whose head seems too small for
its body. Some people prefer the name wapiti, an Algonquin term.
Europeans and New Zealanders know it as red deer, or sometimes
European stag. The scientific name is *Cervus elaphus*. Or is it *Cervus
canadensis*? All these names refer to the same animal – or do they?

Confusion about what, exactly, is an elk started early. The Jesuit
missionaries, the first Europeans to traverse much of eastern
North America, called them elk, but they also used the same word

for moose or even buffalo. The seventy-three-volume *Jesuit Relations*, which consists of their letters back to France, lists elk in its index but follows the entry with the early scientific name for moose, *Alces Americana*.[1] The Jesuits were busy saving souls, not studying wildlife.

Champlain did better. Elk, as we know them, he called cerfs, in rough conformity to the scientific name of the European stag or red deer. He managed to distinguish elk from moose, calling moose by the French word *orignac*, which originated with Basque fishermen who preceded both him and the Jesuits to the east coast of Canada.[2] In the days of early settlement, New Englanders called elk grey moose, as distinct from the black moose, which actually was a moose.[3] Maybe the shade of animal varied with the potency of the moonshine.

The taxonomic confusion over elk is still with us today. Now, however, the question of what is an elk introduces a whole new stratum to our concept of "species," as well as to our understanding of adaptations and their inheritance – for plants, animals, and us.

One cold, starlit autumn night in Colorado's Rocky Mountain National Park, we lay in our sleeping bags and listened to elk music echo and re-echo off the mountainsides as males proclaimed their ownership of harems. Their deep chest acts as a bellows, and the bugling cries made the mountains ring. Ernest Thompson Seton called it "the most inspiring music in nature. Here is this magnificent creature, nearly half a ton in weight, strong as a bull, fierce as a lion, in all the glory of his new antlers . . . 'I am out to fight,' he roars, in tones that tell of his huge round chest. . . . 'On this fight I will stake my range, my family, my social position, my limbs, my life.'"[4]

Early the next morning, before the sunlight atomized the frost, there in a silver meadow stood a big bull with antlers reaching for the sky. Rocky Mountain elk are impressively large creatures with antlers that spread up to 1.3 metres (more than 4 feet). Although the extinct Pleistocene Irish elk would have made him look small, in today's world he is a giant.

Eons of surviving open country attacks by big, fast predators made him so. Since 1995, in the years following wolf reintroduction into Yellowstone National Park, we have witnessed many dramatic attacks on elk. Initially, wolf packs numbering up to 25 animals routinely ran into herds of 200 and 300 elk, scattering them in panic. The elk were naive; no wolves had been present for sixty years. Now the elk have smartened up and stay in smaller, alert herds. Regardless, they still reflect their evolutionary acquaintance with fast-running cats, like the sabre-tooth and the North American lion, as well as with wolves. On level or downsloping ground, elk commonly leave the wolves panting far behind. Size, strength, and stamina characterize Rocky Mountain elk, which are well adapted to grassland, shrub-steppe, and open forest environments.

Not so the Roosevelt elk, denizen of west coast rainforests. A typical encounter with them happens in heavy cover, which is where we found them one rainy April day while backpacking in Washington State's Olympic National Park. The Hoh River Valley features huge Sitka spruce trees, cloaked with mosses and lichens, and is home to Roosevelt elk. Sunlight filters down only weakly; rain seeps through in a steady fall. We spotted their phantom-like forms out on a river flat choked with vine maples and alders, feeling their way along – slowly, carefully, cautiously. Smaller in size (although weights sometimes are comparable), with a darker coat and skinnier antlers, they were different from Rocky Mountain elk in both appearance and behaviour. Stealth, not strength, provides their defence against predators, which is suitable in the forest. They appear as shadows among the trees, where their smaller racks of antlers allow manoeuvrability.

Phantom-like, too, are the Tule or dwarf elk of southern California. Remnants of past greater abundance, they live in semi-desert environments near remaining wetlands, where tall grasses and shrubs provide cover. We caught a glimpse of them one evening in a wildlife refuge on the San Joaquin River. As the sun was setting, three females emerged from the marsh grass, alert, hesitant, watchful. They must have disliked what they saw – maybe us in the distance – because they turned and vanished back into the marsh.

Based on the physical differences among these elk, taxonomists place them in separate subspecies: *nelsoni, roosevelti,* and *nannodes,* respectively. They constitute three of six subspecies of elk in North America, all distinctive enough to have been recognized by taxonomists since the early 1900s. This distinctiveness most probably arose through isolation from one another during the Wisconsinan era, the last 60,000 years of Pleistocene glaciations.[5] As the product of isolation, the differences in appearance have long been assumed to be genetic.

Wrongly so, believes ungulate and animal behaviour specialist Valerius Geist. The author of an authoritative book, *The Deer of the World,*[6] his hypothesis for elk is novel. Geist was an originator of novel hypotheses even back when he and I (John) shared an office as Ph.D. students at the University of British Columbia. He had read that the way to thwart aggression was to counter it with aggression. During his student days, an opportunity to test that hypothesis presented itself when he was a flagman on a lonely stretch of highway in northern Alberta. While standing there by himself, he noticed a bull off in the distance that had managed to get through a fence and walk up onto the road. The bull at first just sauntered in his direction, but then picked up speed. With no trees nearby, Val had two options. One was to run away, which seemed like a losing choice, because he could not outrun the bull. The other was to call its bluff, if it was a bluff, and charge. He chose the latter, and ran at the bull. As the distance between them rapidly decreased, it became obvious that the bull was not bluffing. Exercising prudence, Val abandoned the experiment and veered off in an arc into the ditch and over the fence. The bull stopped, appeared perplexed, and walked away.

His hypothesis regarding elk is not as life-threatening. He believes that most North American subspecies are not subspecies with genetic differences at all but simply ecotypes. The differences in appearance are caused by the environment, not differences in genetic makeup. Rocky Mountain elk display what he calls a "luxury phenotype" (equate the word "phenotype" with "appearance"). They live in a land of relative resource abundance. In contrast, Roosevelt and Tule elk, which live in lands of relative resource shortage, exhibit a "maintenance phenotype." All three are adapted to the environments they inhabit. Although some research has concluded that genetic differences exist among the subspecies,[7] Geist maintained that if you put them in the same captive environment and fed them the same food, any differences in their offspring would be unnoticeable.

The novelty of this interpretation is that it gives the environment an added role outside its accepted one of selecting the proportion of different genes in a population. The viewpoint that the presence or absence of genes alone determines the appearance of a species is changing. Helping foster that change has been the difficulty of the Human Genome Project in meeting popular expectations and ushering in a new era of genetic engineering and gene therapy. The project, completed in 2005, involved mapping the function of all the human genes – about 22,000, it turned out. The hope was that knowledge of the genetic sites of various human frailties and defects would provide a first step to overcoming them. While considerable ongoing research successfully exploits this new knowledge, genes introduced therapeutically into new individuals and expected to perform in a certain way have often failed to operate. The problem? More is involved than the simple presence or absence of genes.

Hints that this was the case have been around for a long time. The "nature or nurture" (inheritance versus environment) debate dominated biology, particularly the fields of animal behaviour and early physical development, in the mid twentieth century. It was abandoned when it became apparent that the choice was more complex than one or the other. Instead, nature and nurture overlap. For example, an organism can inherit a propensity to learn. Furthermore, separation of nature and nurture is not testable in any practical terms, so science hit a brick wall. This topic has remained a grey area in biology ever since.

Unexpectedly, we confronted this grey area in our student days studying ptarmigan. We raised ptarmigan chicks through several summers at the University of Alaska, after which some were autopsied for the sake of science to look at the manifestations of stress. While dissecting the birds, however, we noticed a remarkable thing. They all had gallbladders. This discovery might seem trivial, but members of the Tetraonidae, or grouse family, are not supposed to have gallbladders. A gallbladder is a thin-walled sac situated along the cystic duct between the liver and the small intestine that collects bile made by the liver. It is not a vital organ, as it operates only to aid in the metabolism of fat by timing the release of bile to the presence of food. The wild ptarmigan out on the tundra, from the same stock as the eggs, did not have gall-bladders. Somehow, captivity had induced their formation.

It's likely that no animal physiologist thinks of a gallbladder in any species as anything but a genetically endowed organ. Yet here was evidence to the contrary. We speculated on its cause, and finally linked it, belatedly, to the diet we were feeding these young birds, which was considerably higher in fat than their wild diet would have been. We had laboriously chopped up and fed them chicken eggs, and that, it turned out, was the source of the high fat. The ptarmigan, in response, had developed a gallbladder to help digest it.

For years thereafter we told the story as a bit of a joke, saying that we never tried to publish this finding because there was no "Journal of Scientific Trivia." In retrospect, this observation was possibly more important than the results of the Ph.D. thesis. It fits the evidence emerging today that the environment has a more direct influence on organisms than just simple gene selection. It is evidence of what biologists call "phenotypic plasticity," as was Geist's interpretation of the differences in the elk.

The term "phenotypic plasticity" has been around for a long time but is the focus of renewed interest. It has at least two meanings, one classical and one more recent and expanded. Both share the concept that the phenotype of an individual is the one-to-one physical expression of its genes. If phenotype is a cake, then genotype is the recipe. Consequently, a population with a lot of genetic variation will have a lot of individuals that look different – that is, the population will have phenotypic variation or plasticity.

The expansion of this interpretation today is that even populations *with the same average complement of genes* may have different average phenotypes. In other words, working from the same recipe, two cooks bake up different cakes. That possibility explains the ptarmigan case, because the genetic makeup of the wild birds with no gallbladders and the captive ones with gallbladders was the same.

How can the cakes be different? The explanation has two parts. One is that genes rarely work as individuals; they function in interconnected networks – gangs, if you like. As explained in Chapter 1, Gregor Mendel's discovery of genetic variation in 1866 was based on an uncommon occurrence, as it turns out: one gene responsible for one genetic trait. More commonly, members of gene-gangs influence the expression of other members – turn them on or off. Such regulatory genes play a vital role in the development and maintenance of organisms.

Unravelling these networks in the Human Genome Project and related endeavours has not been easy. But this complication was expected. After all, in a developing embryo, all cells have exactly the same genetic makeup. Yet some cells differentiate into liver, others into heart, others into muscle. They are able to do so because of gene regulation, "switchboxes" that allow different genes to be expressed or repressed during development. We still have a great deal to learn about genetic switchboxes and the actions of gene networks.

The other part of the explanation addresses the question of what regulates the regulators, and that subject is even less well understood. Increasingly, the answer seems to involve the cell's internal biochemical environment. Enter epigenetics, which means "beyond genetics," a new dimension in heritability, or the ability of a characteristic to be passed on to subsequent generations.

The term epigenetics has also been around for a while, since the early 1940s. Initially it was applied to the differentiation of cells into various tissues and organs in a developing embryo. Even though this was an active area of research, the word itself was little used for the first five decades of its existence. Then, suddenly, it resurfaced, bursting onto the scene with new meaning. It became clear to cellular biologists that the mechanism that caused different types of cells to develop was "beyond genes" yet was inherited. Thus, epigenetic inheritance came to mean the transmission of information to subsequent generations that does *not* involve changes in DNA (genes).[8] But every student of biology has been brought up to believe that genes are everything in inheritance. This new concept sounded like Lamarckism – the inheritance of acquired characteristics – an idea believed to be quite dead for almost a century. The popular press picked it up in that way.[9]

Not only was the popular press excited. In 2005 an article appeared in the research journal of McGill University heralding "The Birth of a New Science," reporting on the research of Moshe Szyf and Michael Meaney. They had demonstrated that the expression of individual genes can be permanently altered by such seemingly innocuous influences as diet or how others treat us – in short, the environment, or, put another way, nurture affecting nature.[10] "Epigenetics," Szyf said, "will completely change the face of medicine."

Examples that have shown up include the classic case that is nearly always cited of agouti mice. Mice with a particular gene show several traits: yellow coat instead of brown; fat bodies; and tendencies towards cancer and diabetes. Offspring that inherit this gene exhibit the same traits. Researchers[11] managed to "silence" the gene with diet, specifically foods rich in methyl donors. The methyl molecule was able to attach itself to the gene and turn off its expression. Moreover, offspring of these "methylated mice" also inherited the methyl donors and turned out to be normal mice.

Other interesting examples that involve inheritance in subsequent generations include evidence that the sons of men who began smoking before puberty were more prone to obesity.[12] And rats exposed to the common pesticide methoxychlor (classed as non-toxic by the U.S. Environmental Protection Agency) produced four generations with reduced male fertility.[13]

What was happening inside cells to accomplish this sort of feedback from the environment? Among a variety of mechanisms is a behavioural one: rats that received healthy doses of maternal licking as pups grew up to be calmer and more socially adept than those with less attentive mothers. The authors of this study concluded that maternal grooming brought about a chemical change in the mechanism of the brain that regulates stress hormones in the offspring, and that change persisted throughout life.[14]

The involvement of stress in influencing inherited characteristics has received considerable attention, especially in human medicine and mental health, but it has relevance to the variation in species – the stuff of evolution and ecology – too. Here, the work of Dmitry Belyaev, who worked at the Soviet Academy of Sciences, has come to the fore. In the late 1950s he started an experiment, which was continued even after his death in 1985, that involved selection for tameness in the silver fox. He selected and bred the individuals in each generation that were most docile. Today, the foxes are quite dog-like in behaviour, eager to please their human handlers and compete for their attention.

That was not all. In fewer than twenty generations, levels of reproductive and stress hormones had changed, and the foxes developed physical features such as drooping ears and tails, white spots on their fur, shorter legs and tails, and different skull shapes. These new phenotypes appeared too frequently to be the result of new mutations. In 1979, Belyaev explained his results with surprising accuracy in view of what is known today. He attributed the different phenotypes to the consequence of epigenetic rather than genetic change. Selection for tameness had altered the hormonal state of the foxes, which in turn had affected part of their cell biochemistry and thus activated many normally silent genes in both the body and the inherited genome. All these new features were the consequence of the stress associated with tameness, which caused hormonal changes that revealed previously hidden genes and made them available for selection.[15]

Where all this rings a bell with ecologists is in how it helps explain historic patterns of evolution. Bursts of new species have appeared in the fossil record immediately after each of the five great extinction events on Earth. Repeatedly, out of the ashes of species annihilation have come arrays of biological novelty.[16] The normal explanation for this phenomenon is the increased opportunities for surviving species to capitalize on reduced competition and spread into new niches, forming new species in the process. This explanation may be valid, but it fails to explain where the necessary underlying genes came from. Now, these bursts of new species may also be recognized as the result of altered environmental influences on existing genes. Adding to the likelihood of this explanation is that these post-extinction times have been ones of severe environmental stress, shown by Belyaev's foxes to be an epigenetic trigger.

Is epigenetic inheritance a vindication of Lamarckism? To the extent that it represents changes in individuals as a result of environmental influences, it does indeed fit the notion of inheritance of acquired characteristics. Epigenetics involves a feedback from the environment that may result in adaptations that are heritable. But, diverging from Lamarck's explanation, an individual's interaction with the environment does not make new genes. Rather, the environment may only call up for expression hidden genes that are already there.[17]

Where did these hidden genes come from? Over time, silent mutations accumulate. Most of these mutations are not immediately beneficial, and if not detrimental, they will persist in their unexpressed form. Consequently, most organisms possess an extensive bank of genes that normally are not expressed. Then, an environmental condition may change and trigger their expression.

Abetting the accumulation of hidden genes, too, is the situation that most physical traits in organisms are the result of gene systems, not just individual genes. These systems are well buffered against change, and most mutations are not expressed. That is why generations of any lineage persist for a considerable time in an almost identical form. When conditions change, however, epigenetics may kick in – a streamlined way for the environment to call up the expression of silent genes, and in that way improve the chance of a beneficial one being expressed.

Thus, quick adaptation to environmental change does not depend solely on entirely new mutations, or on the recombination of existing genes, as was once thought. It may depend more on epigenetic change that unveils genetic variants already present in the population.

And the ptarmigan's gallbladder? Most likely the high level of fat induced a biochemical change that turned on a gene for the organ's development, or maybe turned off a gene that inhibited its development. Either way, it stands as a likely case of epigenetics.

Ron Nowak is a quiet, unassuming man, retired from a career as a taxonomist in the U.S. Fish and Wildlife Service. His career was based largely on measuring bones, and he is the undisputed expert in that regard for the wolf. He has searched the museums and universities of North America for specimens, ancient and present-day, and from them has traced out how the gray wolf, *Canis lupus*, most likely swirled into existence over the past million or so years.

Claiming a more recent but accepted approach to taxonomy, too, is California geneticist Robert Wayne. His bold, assertive personality is the opposite of Nowak's. The two men differ, as well, in their conclusions about wolf taxonomy.

In 1995, Nowak published what has been generally accepted as a five subspecies classification of the gray wolf in North America based on differences and discontinuities in bone measurements.[18] Four of these subspecies make sense based on hypothesized isolation during the ice ages in ice-free refugia that allowed for genetic divergence. The fifth subspecies, *Canis lupus lycaon*, the gray wolf of the eastern forests, from central Ontario and Quebec south to North Carolina and Tennessee, was an enigma, with no obvious glacial refugium. In addition to gray wolves, Nowak described a second full species of wolf in North America, *Canis rufus*, the red wolf that once inhabited the southeast – Louisiana, Arkansas, the Carolinas, and east Texas.

Wayne disagreed about the red wolf. Based on an analysis of

mitochondrial DNA, he concluded that the red wolf was nothing more than a gray wolf–coyote hybrid.[19]

Initially, this disagreement about the red wolf was not relevant to our wolf research in Algonquin Park, taking place far to the north of red wolf range. Both biologists agreed that in Algonquin we were studying the *lycaon* subspecies of gray wolf. However, both wanted information from us to help them develop their continental pictures. From the 150 wolves we captured and radio-collared, we took body measurements that we sent to Nowak and drew blood that we shipped to Wayne for genetic analysis.

These shipments to California caused us problems. The blood had to be kept frozen, so we packed the vials in dry ice. But border inspectors are leery of biological material like that, and because wolves were an endangered species in the United States, and blood is considered a "body part," our statement of scientific purpose prompted some scrutiny. One Monday morning we received a phone call from U.S. security in Tennessee, which for some unknown reason had received the vials addressed from Ontario to California. The vials had arrived on Friday but the weekend had intervened. Despite the warnings on the package, nobody thought to refrigerate them. The security officer, however, seemed sure he had lucked into a criminal activity of some considerable magnitude, and telling him that the blood was for genetic analysis made him even more certain. However, as the interrogation proceeded, he gradually shifted his interpretation from subterfuge to meaningless academic activity and terminated the cross-examination. When the vials finally reached the University of California, the dry ice was long gone.

After that we switched to a genetics lab closer to home and began to work with Brad White at McMaster University in

Hamilton, Ontario. The results that came from his lab were astounding. The genetic profiles of Algonquin wolves matched very closely those of the red wolf, not the gray wolf. There appeared to be no genetically distinctive *lycaon* subspecies of gray wolf at all. Nowak was intrigued, because here was an explanation for his enigma of *lycaon* having no glacial refugium. None had been needed. *Lycaon* was really the red wolf that had likely lived south of the ice in the southeastern United States during glacial times.

Brad White, Paul Wilson, and their colleagues went on to publish a proposal that the red wolf and the *lycaon* subspecies of gray wolf should be classified together.[20] We sided with their interpretation rather than with Wayne's, because it was based on a broader sampling of DNA (nuclear as well as mitochondrial). Another reason was that for years after giving talks and showing slides of Algonquin wolves, any red wolf researchers in the audience would invariably comment that we were showing their wolves. Both body size and coat colour were good matches, as were characteristics of the face and ears.

Then, in 2002, Nowak complicated the issue. He published a re-evaluation of his bone measurements and stated that the *lycaon* subspecies of gray wolf and the red wolf were indeed different, as he had originally thought.[21] He acknowledged that his conclusions clashed with the genetic interpretation of White and colleagues, and left it at that. Left, too, was an underlying sense that one or the other interpretation must be wrong.

However, both may be correct. Because epigenetically derived phenotypic plasticity can create physical differences from the same genotype, you could indeed have physical differences between the red wolf and the *lycaon* subspecies of gray wolf that

Nowak identified, with no genetic differences that White and colleagues identified – just as it is with the different ecotypes of elk.

This conclusion would not surprise Val Geist, who believes there is so much phenotypic plasticity built into species, especially the hoofed mammals he has studied, that taxonomy based on body characteristics is relatively meaningless. But wait. That depends on what we are trying to understand. If the purpose of taxonomy is to determine lineages and to resurrect past evolutionary relationships, then genetics does the job. Genotype is paramount. However, if the purpose of taxonomy is to classify biological diversity, then different ecotypes, with their different environments, regional adaptations, and histories, must be the overriding criterion. Phenotype is paramount.

If that conclusion doesn't set back traditional taxonomy based on physical appearance far enough, Geist has gone further to state that, "The phenotype [meaning here the range of physical variation in a species] is the genotype's defence against environmental change, against evolution, and against extinction,"[22] an observation that other biologists have made, too.[23] This statement sounds extreme, but it is correct because of epigenetics. In the face of environmental change, an epigenetic response will be most immediate, with cell biochemistry calling on hidden genes. Only if suitable adaptation does not then occur will genetic change via mutation be an option.

Wolves, like hoofed mammals, exhibit plenty of physical variation. The high arctic subspecies *arctos* is predominantly white; wolves in northern Alberta and British Columbia of subspecies *nubilus* are both grey and jet black; Mexican wolves of the subspecies *baylei* and the red wolf are reddish behind the ears and on the legs. Arctic wolves have short ears and bulky bodies; Mexican wolves have long ears and a lean physique. How much of this variation is genetic and how much is epigenetic awaits further analysis. It is not easy to separate these entwined fields. Biologists Eva Jablonka and Marion Lamb state, "We believe that it is highly likely that there is much more epigenetic inheritance [in species in general] than has so far been identified. People have hardly started looking for it."[24]

As pronounced as the physical differences are in wolves, they are nothing compared with what humans have managed to select out of the wolf pool of hidden phenotypic plasticity. All domestic dogs come from wolves, and in no other species has the latent pool of available variation been so exposed. Humans have not caused any new genes, merely triggered many existing ones through selective breeding. What Belyaev did with silver foxes, dog breeders have done with each divergent line of domesticated wolf, resulting in the impressive array of dog phenotypes – a dog type to complement every human personality type. Wolves, indeed, possess an astonishing phenotypic plasticity that we have managed to reveal.

On an abandoned logging road twenty-five kilometres south of Algonquin Park, we waited for dark one evening amid the hum of mosquitoes and the clatter of late-season mink frogs. The forest around us was not much different from parts of Algonquin Park – rolling hills of sugar and red maple, yellow birch, and aspen. Like Algonquin, the lakes, bogs, and rivers were sketched in with conifers, and like Algonquin, too, it was deer, moose, and beaver country.

When it was suitably dark, we gave a howl, part of our normal census routine along old logging roads, and were answered immediately by a – wolf? The answering howl was high-pitched, an octave or more above normal wolf range. It lacked any deep-chest resonance. It was yippy at the beginning and featured a few barks between howls. That was a shock. We recorded it.

We were there that night in the Madawaska Highlands because of misgivings over the movements of some of our radio-collared Algonquin wolves in and out of the park, and a high level of human killing when they were outside. We expected to hear resident wolves. Instead, the howls were more like those of coyotes.

We were already concerned about coyotes. Brad White and his colleagues had found some coyote genes in our Algonquin wolves, indicating hybridization. This discovery had helped confirm that the Algonquin wolves were red wolves, because red wolves are known to interbreed with coyotes whereas gray wolves do not. We did not, however, expect to find coyotes living in closed forests so close to the park.

Hybridization alarmed us, because the red wolf became officially extinct in the wild in 1970 due to "gene swamping" from coyotes. In the early 1900s, as coyotes expanded eastward into wolf ranges from their original homeland in the central plains, they interbred with remnant wolf populations. Wolves grew progressively smaller and became more coyote-like in appearance. (Since then, a captive breeding program has resulted in the release of red wolves in a wildlife refuge in North Carolina, and the population there is doing reasonably well, except for the continuing problem of coyote interbreeding.)

With a concern over hybridization, our field crews moved in and radio-collared five of these Madawaska animals. Their measurements were small, intermediate between wolf and coyote but tending towards the latter. The crews drew blood, too, which was duly sent off to Brad White at McMaster University. The results that came back put these animals genetically closer to New Brunswick coyotes than to Algonquin wolves. Apparently, the wolves that had once inhabited this forested region had been "gene swamped" by coyotes too, just like the southern red wolves.

Coyotes are relatively new to Ontario; they expanded their range across the southern part of the province only in the early 1900s. The gene swamping had taken place since then, facilitated, we concluded, by the killing and subsequent fragmentation of the wolf population, just as had happened in the United States. The next winter we collected canid carcasses from trappers in the area and found that close to half the population was less than two years old, evidence of a high level of human exploitation.

While the animals from south of Algonquin Park had more coyote genes than most of the wolves in the park itself, five animals that we collared in the heart of the park were small, like the Madawaska ones.[25] Obviously, the same gene swamping was also beginning to happen even within the park. If not stopped, it threatened the existence of wolves – red wolves no less, a new species for Canada. After much publicity and considerable public conflict, the Ontario government permanently banned wolf killing all around Algonquin Park. Coyotes were included too, because traps and neck snares do not discriminate, and because in intermediate form they look so similar.

South of the park, however, is a radically different environment, despite the superficial similarities. Humans make the difference. Here are numerous logging roads open to the public, snowmobile and ATV trails, small towns and farms. In this cutover, shot up, trapped, human-inhabited landscape, what sort of canid would the environment favour? Remember that wolves

have a wide, hidden phenotypic plasticity to draw on, epigenetically and genetically, enhanced in this case by hybridization. The
environment would not call up a "luxury phenotype," as Geist
described for Rocky Mountain elk, or even a "maintenance phenotype," like that of the Roosevelt elk, because conditions
included not only a relatively short supply of ungulate prey but
heavy exploitation and a need for the animals to exercise extreme
caution. Any scent beside a trail could lead to a trap, any downed
tree to duck under could conceal a snare. Best adapted would be
exactly what was currently there: an intermediate-sized animal
capable of killing deer and beaver just like wolves – maybe not
moose, but they were less abundant there – and able to subsist
more successfully on hares, mice, and other small mammals, just
like coyotes. Favoured, too, would be an animal intermediate in
social behaviour between wolf-like reliance on the pack and
coyote-like solitariness. Packs come with the cost of being conspicuous to humans but the benefit of being better at killing large
prey; solitariness the reverse. Tending to overcome a high level
of mortality and to persist, too, would be an animal with limited
social restraints on breeding and inclined to have its first litter as
a yearling, like coyotes. Behold, a new canid ecotype – gene selection and epigenetics – evolution in action.

What, if anything, does this expanded understanding of the various
sources of phenotypic plasticity mean? Will it shake up the world?
It could, but it is too early to know. If a Human Epigenetics Project
gets underway, or if research on the implications of epigenetics to
cancer or other aspects of human medicine bears fruit – and there
is a lot of interest in that – then a few decades from now we may
be viewed as having lived in the medical Dark Ages. Referring to

all species, Jablonka and Lamb wrote, "Our basic claim is that biological thinking about heredity and evolution is undergoing a revolutionary change."[26]

This new understanding cannot help but have implications for conservation. At present, the major focus of species conservation has been on genetics, which is still mistakenly seen by many biologists as the sole basis of heritability and thus of distinctive species. As environments alter rapidly, which is a predicted consequence of global climate change, species adaptations through epigenetics may prove more important than adaptations through genetics. Evidence shows that variation through epigenetics is faster than through gene selection alone, especially if linked to environmental stress. Living things with rich epigenetic systems may exhibit the greatest phenotypic plasticity to meet the conditions of the near future. A new field, epigenetic ecology, may be vital if we are to find ways to maintain biodiversity.

As a starting point, shouldn't we be preserving as much as possible of the phenotypic variation that species exhibit – subspecies, regional populations, any variant populations, wherever they exist? Doesn't that make ecological sense as a hedge against the future, since epigenetics emphasizes again the close link between species and their environments? They are integral and inseparable, one and the same.

And so, Rocky Mountain elk in their environment and Roosevelt elk in theirs are different entities, even though genetically indistinct. So are the subspecies – likely ecotypes – of wolves. Even if we cannot attach adaptive significance to all the physical differences, they still are evolutionary products in their own right. But in a changing world, who knows what variants will emerge as the best fit?

Then there is the satisfaction of knowing that we are one step closer to recognizing how so many marvellous adaptations occur. There is indeed environmental feedback beyond that provided by gene selection alone, and if conditions are right, at times it may work expeditiously. Lamarck was partially correct. He was a fore-runner of an important component of self-organization. In years to come we will understand that concept better.

CHAPTER THREE

DAWN CHORUS

Birdsong – meant to spread joy in human hearts, music to enrich the soul? Not so! Birds sang for some 150 million years before human ears existed to hear them or human minds to be enthralled. Birds don't really give a damn about us. Birdsong is for the birds. But now, after having been studied from two aspects – behavioural and physiological – new meaning is emerging from birdsong that provides an in-depth look into the intricacies of adaptations, the very stuff of biological success.

BIRDSONG, ALWAYS BACKGROUND to our spring and summer fieldwork, took on dramatic new meaning for us in 2000 after we moved to British Columbia. Then, comprehending its intricacies became a focus of our research. We wanted to test the use of song as a behavioural tool in making species distinctions, along with the conventional methods of genetics and physical appearance. A thorough understanding of the diversity of life, which is important to conservation, depends on being able to classify distinct species, or gradations between them, as the fundamental players in the game of life.

Soon we found out that because birdsong is so intricate, the research was in a state of some disarray. So many birds have such rich repertoires of song that the vocal variety seems excessive. Ornithologists, in searching for explanations, have tried to classify song in various ways: learned versus hard-wired, melodious versus noise, complex phrases versus simple songs. The classifications have not spawned any very insightful hypotheses. It seems as though evolution declared a free-for-all with birdsong, with no rules, no restrictions.

But evolution does not take time off. Anything so prominent in biology as the variety in birdsong must be adaptive in important ways. Frighteningly, the time left to find answers may be shrinking, because in North America many bird populations are declining. Do we want to live in a birdless world? Witness the titles of various books: *Silent Spring* (1962),[1] *Where Have All the Birds Gone?* (1989),[2] *Silence of the Songbirds* (2007).[3] Can it come to that?

We humans interpret birdsong as beautiful because at times it seems to incorporate features of our own music. Music is comprised of notes with regular, mathematically related harmonic overtones, which is lacking in noise. Music places notes into a juxtaposition of phrases, rhythm, and cadence. Once, it was popular to analyze birdsong by keys, tones, half-tones, octaves, and other human constructs, but birds do not operate by human musical scores. They can spill out notes in a complexity that puts human singing to shame.

Not all birds sing. Songbirds are just one group of birds, the oscines, represented by about 4,600 species. The oscines lie within the giant order Passeriformes, or perching birds, that make up about half the species of birds in the world. The other half of the world's birds, the non-passerines, are non-musical and fit within other orders that include ducks, hawks, grouse, cranes, and shore-birds. And among songbirds, the males do most of the singing.

Also within Passeriformes are another 1,100 species of sub-oscines that make noise instead of music, although there are exceptions in both cases. A cedar waxwing's high, seedy sound hardly qualifies as musical, although it is an oscine, and conversely, a bobwhite's call sounds musical even though it is a suboscine. Anatomically, however, the two groups possess very different mechanisms for producing sound.

It was a nondescript little bird with an olive-green back and white belly, a Cassin's vireo, that started us studying birdsong. In the Okanagan's forests of ponderosa pines and Douglas firs, we kept hearing a song that sounded like a solitary vireo, familiar to us from spring days in the eastern hardwood forests. This western bird sang with the same spaced out, chunky-sounding phrases. However, in a new edition of a field guide to western birds we noticed that the solitary vireo was not listed, although it had been in earlier editions. Instead, the bird with almost identical plumage and song was called a Cassin's vireo. Previously, Cassin's vireo had been described as a western subspecies of the solitary vireo, but in 1994 full species status was conferred to it. The taxonomic split-ters had been at it again, using genetic evidence to make two species out of one.

Cassin's vireo is not the only western species masquerading as a familiar eastern one. So is the bright little Audubon's warbler, which differs from the eastern myrtle warbler by sporting a yellow throat rather than white. Here, the taxonomists had done the opposite – lumped them together, despite their obvious differences in plumage, and given them both the same name: yellow-rumped warbler.

Other east-west counterpart species differ more substantially in plumage but very little in song: black-headed grosbeak in the west and rose-breasted grosbeak in the east; western tanager and scarlet tanager; lazuli bunting and indigo bunting. This west-east dichotomy is a distinctive feature of North American passerine birds, the result of many species being wedged apart during continental glaciations that began as early as 5 million years ago.[4] Separated, over time they diverged in physical characteristics and/or song, to become what sometimes are listed as different species, such as the vireos, and other times are listed only as subspecies, such as the yellow-rumped warbler. The times of separation undoubtedly varied, so that today some pairs of counterpart species are more similar to each other than others.

This situation has caused a lot of confusion. The keeper of bird records, the American Ornithological Union, has lumped, split, lumped, and split species as new evidence has shown up. Today, genetic evidence trumps plumage as the key diagnostic feature used to distinguish species. Whether it should be accorded that overriding importance, however, is debatable. We now know that locally adapted ecotypes may be the consequence of epigenetic, not genetic, inheritance. We also know that phenotypes (physical appearances and behaviours), not only genotypes, are relevant for conservation.

So, somewhat naively, we began to explore song as a criterion of species identity, something that birds evaluate from their own perspective. With a minidisc player and battery-powered speaker we started playing the songs of eastern birds to their western counterparts, and vice versa. We developed a protocol of playing the counterpart's song for two minutes, noting the bird's behaviour, then playing its own species' song. Sometimes we reversed the order to avoid experimental bias. As a control, we often played a third song of a species in the same taxonomic grouping but not so closely related.

The range of responses was dramatic. At one extreme was a high level of aggression: the bird became very agitated, repeatedly flying low over our heads and landing momentarily a few metres away. Intermediate were various categories of exploratory behaviour: the bird stopped singing and approached as if to satisfy its curiosity. At the other extreme was no reaction whatsoever: the bird just kept singing or flew to another song perch nearby. Responses were temporary, waning after about five minutes. We returned on successive days and are confident that our experiments had no lasting effect.

Cassin's and solitary vireos, our initial subjects, were especially interesting because where their ranges overlap, their similarity in plumage could lead to hybridization from mistaken identities. Would song discrimination act as a barrier? Apparently not for Cassin's vireos, which turned out to be just as confused as we were. In eight experiments, only once did a bird respond more strongly to its own species. Four birds responded more aggressively to solitary vireos, the wrong species (if indeed the American Ornithological Union is right and they really are different species). Two birds showed the same strength of response to both species, and one bird responded to neither.

In the eastern forests near Algonquin Park we tested solitary vireos and found better discrimination. In twelve tests, solitaries responded most often to their own species seven times, to Cassin's once, and showed no interest in either species four times.

We have no hypothesis to explain these differences. However, after listening to them sing so many times, we can now consistently outperform the Cassin's poor ear. Its song has a slightly burry quality to its phrases, whereas the solitary's notes are purer.

Because Cassin's vireos do so poorly in distinguishing their own species from solitary vireos, song for them cannot provide any barrier against hybridization. The only barrier seems to be the limited overlap in geographic ranges. In an admission of uncertainty, the *Sibley Guide to Birds* says that these vireo species "are rather poorly differentiated. Variation in song often follows a cline [transition across a distance], and individuals can learn the 'wrong' song. Intermediate birds (and perhaps hybrids) should be expected, and not every individual will be identifiable."[5]

Insect specialists encounter hybridization more commonly than vertebrate biologists do and sometimes speak of "species swarms" where distinguishing discrete species among many variants is arbitrary and therefore invalid. Ornithologists, too, are beginning to accept the similar idea of "superspecies groups" – some of the gulls, for example. Acknowledging groupings such as these may be more valid than trying to shoehorn everything into discrete species compartments.

Our playback experiments are admittedly somewhat exploratory. We still have not hit on neat, testable hypotheses, but other researchers have had the same problem. Much of the literature on birdsong is descriptive, not analytical; it addresses what is happening, not why. Part of the problem arises from confounding variations in song.

It took a few Swainson's thrushes to show us that. We downloaded the recording of a float trip on South River near Algonquin Park to our computer, then isolated a Swainson's thrush that had obligingly sat in a riverbank pine and poured his music into our microphone. Then we slowed down the song to one-quarter speed and made a sonogram of it – a graph with frequency in kilocycles per second on the vertical axis and time on the horizontal. On the sonogram we could see the variations in pitch and phraseology, and hear them simultaneously, too. Swainson's sings up the scale in ascending phrases, and as beautiful as its song is, when you slow it down its full complexity is revealed. Hidden phrases, trills, arpeggios; what sounds like a steady note may actually be a trill, a slur, or a chord.

We picked the next song the thrush sang, only about three seconds later, downloaded it, slowed it down, and made a sonogram. The songs were not the same, even though when played at full speed they seemed to be. The character and even the number of phrases were different. The same was true for the next song, and the next. That one bird, it turned out, had a repertoire of six different songs.

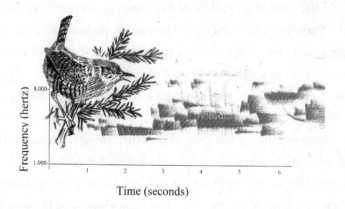

Time (seconds)

Most songbirds, like the Swainson's thrush, sing more than one song. Song sparrows have a repertoire of about ten songs, wood thrushes ten to twenty, robins around fifty, and the greatest minstrels of all, brown thrashers, have over two thousand. However, some birds sing only one song, such as indigo buntings, chipping sparrows, ovenbirds, veeries, and white-throated sparrows.[6] Some vary the length of songs by sometimes cutting them short; Baltimore orioles and fox sparrows do that. Swainson's thrushes sing their repertoire in a relatively fixed order; lark sparrows never repeat two songs in a row; brown thrashers always repeat two songs in a row; western meadowlarks sing the same song repeatedly before turning to another.

This phenomenon of multi-song repertoires in most species poses a problem. When conducting a playback experiment, could our results be influenced by what song type we play? Does it matter whether it is one in the subject's repertoire, or a song the bird is used to hearing, or not hearing, from a neighbour? That uncertainty is compounded by another source of variation in birdsong, regional geographic dialects. Swainson's thrushes gave us an example of that, too.

In the dusk of an early summer evening, we stood on the edge of a farm field near the city of Duncan on Vancouver Island. A full moon was just rising above the spindly tops of the firs. Venus punctuated the western sky. A couple of nighthawks flapped erratically overhead. At least six Swainson's thrushes were singing from various places around us.

But something was odd about their songs. Instead of one simple, quick note at the beginning, before the notes ascended the scale, there were two. Nowhere else had we heard that. We rushed back to the truck for our recording equipment. The birds sang on in the growing darkness.

We did a quick download, slowed the songs, and examined the sonograms. Yes, they all began with two quick, identical notes. The next morning we recorded another Swainson's thrush singing nearby, and stayed with him to see if there was variation. There was. While most of his songs began with the two notes, now and then, about every tenth song, he slipped in a song that began with only one note. Even fewer, one in twenty, began with an unprecedented three notes.

Since then we have checked out Swainson's thrush songs more broadly on southern Vancouver Island, and once during a canoe trip on nearby Quadra Island. All birds give two-note beginnings almost exclusively. In the Okanagan, however, one-note beginnings are usually the rule, as they are in Ontario and in all the bird recordings we can lay our hands on made anywhere in North America east of the Coast and Cascade mountains. Not invariably, but over 90 percent of the songs begin with only one note.

Here was an example of regional dialect, like different human accents or localized figures of speech. Yet this example paralleled differences in appearance, too, which is not always the case. Swainson's thrushes are described in two distinct groupings, one in the extreme west and the other living all across Canada and parts of the United States east of the Coast and Cascade mountains. So we concluded that some degree of past isolation had resulted not only in minor differences in plumage – western birds are called "russet-backed" and the rest are called "olive-backed" – but also in song.

While the dialect differences in Swainson's thrushes are subtle, the regional dialects of other species can be remarkably different. One morning, on a hike through Nevada's high-elevation piñon pine forests, we heard a song that was new to us. We searched out the singer, and to our surprise found that he was a black-throated gray warbler singing a totally different song (to our ears) than those we knew and had recorded on Vancouver Island. So, out came the playback equipment, and we played the Vancouver Island song to this Nevada bird. He immediately recognized this different song and came zooming in, very perturbed.

Similarly, several Townsend's warblers we tested along a steep, forested mountainside on Vancouver Island came in very aggressively to the completely different-sounding Townsend's song on the recording in *Stokes Field Guide to Bird Songs*. Both Townsend's and black-throated gray warblers are split into two subspecies that live in different areas and display different songs. Yet they readily recognized the dialect of their other subspecies, even though they had never heard it before.

Different dialects need not be widely separated geographically. In only thirty kilometres (nineteen miles) along the California coast, Donald Kroodsma, the guru of birdsong researchers, identified six different dialects of white-crowned sparrows, each in a local grouping. Each group's area was known to the sparrows themselves from subtleties of the terrain.[7]

What is the adaptive significance of all this variation in song among species, among individuals, and within individuals? How and why has it evolved? Much research has been directed at these questions, and it continues, because answers have been hard to come by.

One possibility is that variation relates in some way to speciation, that is, to the processes that originate species or keep them distinct. When we look across all the tests of eastern and western counterpart species we have done, we can see that normally the responses are strongest to a bird's own species, with weaker responses shown towards the counterpart, whereas no responses whatsoever are generated by other species even in the same family. Apparently, some genetically endowed common template, some instinctive memory, triggers recognition of the counterpart species. For those counterparts that have never heard the other's songs, because their ranges do not overlap, this memory could date as far back as a few million years to the time of their divergence, although we do not know how ranges may have shifted over time.

So marked is this generalization about the strength of response that when it fails to occur, we suspect something is wrong. The Cassin's vireo's inability to identify his own song is an enigma, as mentioned. Western red-naped and eastern yellow-bellied sapsuckers respond equally well to each other, and may be candidates for some alteration in distinctive species status even though their plumages are very different. The current classification of winter wren is likely a mistake.

Winter wrens in North America are listed in field guides as one species. They have an odd-shaped breeding range showing a bulb in the northeast joined by a narrow neck lying north of the prairie provinces to another bulb in the northwest. One morning, in a sunlit patch of shadowy interior rainforest in British Columbia's Mount Revelstoke National Park, we played songs to a lively winter wren that came in very aggressively when he heard a western winter wren, but retreated and showed only mild interest when he heard an eastern one. The songs sound different to us, too – the western song was more wheezy. The Revelstoke wren was reacting as if the eastern song was from a counterpart, not his own species. And yes, as expected, genetics research has caught up now and indicated that eastern and western winter wrens indeed should be categorized as two different species.[8]

The possibility that differences in birdsong may provide the necessary breeding isolation for new species to form was put forward in the 1960s by pioneer birdsong researcher Peter Marler.[9] If a female must hear only the appropriate song as a prelude to mating, then any deviation from it might provide just as effective a barrier to mating as does geographical isolation, which normally is considered necessary for the formation of new species. Marler's line of reasoning was as follows: birds raised in captivity have demonstrated that all members of the group "oscine" must learn their songs, and do so as they grow up, principally from their fathers;[10] then, because most males tend to return to their natal area to become breeding birds themselves, local dialects develop; finally, if these birds are favoured by females, the result will be local selection pressures favouring local dialects, and breeding isolation will occur.

The beauty of this explanation is that it provides a plausible mechanism for the abundance of songbird species. Always bothersome has been the necessity to accept the conventional wisdom that actual geographic isolation must occur for species to originate. A very large number of glaciers, or islands, or other geographical barriers must have been needed. With birdsong providing behavioural isolation, however, many species could have diverged without that necessity.

Our dialect tests argue against this line of reasoning. If a bird recognizes and reacts to its own species even when hearing it sing an unfamiliar song, as the black-throated gray and Townsend's warblers did, then isolation through song is not possible. Also damaging to Marler's hypothesis is the work of Donald Kroodsma, who showed that young adult males that have learned their father's song may change their songs to conform to a new dialect if they settle the next year in a different place.[11]

Perhaps the role of variation in birdsong is less relevant to barriers between species than to other candidate roles, such as claiming a piece of land and attracting a mate, the primary function of all birdsong. This dual function was first described by British ornithologist Eliot Howard in his seminal book *Territory in Bird Life*, published in 1920.[12] Supporting his claim, it is known now that song variation serves not only for species and subspecies recognition, but for neighbour recognition, too.

One June morning, high in an Okanagan larch/lodgepole forest, we were able to distinguish the boundaries between two singing Swainson's thrushes. Neither bird seemed the least bit interested in the other, even though they were only about 60 metres (200 feet) apart and clearly within hearing of each other. Their songs were somewhat different, however, and from our repertoire of previous recordings of other Swainson's thrushes we tried to match them as closely as possible. When we played a close facsimile of one bird's song, that bird recognized the difference from what he was used to hearing from his neighbour (even though that song was similar to his own) and came storming in. The other bird, however, heard what he presumably took to be his neighbour's song and ignored it. The reverse response happened when we played a facsimile of the second bird's song. In that dramatic way, each bird displayed an ability to recognize his neighbour.

Other researchers have documented neighbour recognition more thoroughly. For example, in white-throated sparrows, the vocal cue is in the spacing of the triplets at the end of the song, which is specific to individual birds.[13] In indigo buntings, the rhythmic cadence of the pairs of notes is most important.[14] All individual members of both the oscines and suboscines have distinctive song elements,[15] sometimes very subtle ones, that are used for individual recognition by members of their own species.

The usefulness of this ability to "know thy neighbour" lies in being able to resolve territorial issues once, without the constant strife of chasing down what could be new and threatening adjacent singers.[16]

But doesn't a wider repertoire make neighbour identification more difficult? What is the advantage in constantly switching songs? Maybe song variation is a form of "showing off" to a female, just like bright colours and displays. An individual with many songs may be interpreted by a female as vigorous and healthy, a good mate. If so, then sexual selection would operate directly to increase repertoire size. But why, then, do some species have only one song?[17]

Alternatively, variation in song may be evoked by competition from neighbours – a heightened effort to get the message through the surrounding din of song. This proposition is supported by Kroodsma's finding that repertoire size is greatest where populations are most dense.[18] Or, the environment may have an effect if it masks some sounds. The acoustical properties of different environments appear to select different features of song. Most obvious are the loud, jumbled notes of the American dipper, a bird that lives by fast-running, noisy water. Some time in the past, natural selection ratcheted up its volume and variability so it could be heard. Other environments exercise more subtle effects. Various studies have concluded that: open habitats transmit variability in sound frequencies better than closed ones;[19] foliage and tree trunks dampen high-pitched sounds;[20] and, wind tends to mask low frequencies more than high ones.[21] These characteristics may be responsible for the tendencies of high-pitched singers to be treetop species, such as blackpoll, bay-breasted, and blackburnian warblers,[22] or open country species, like horned larks.

But if any of these possible explanations are valid, why don't they apply more consistently? Why is song variation so variable? The answer may lurk in the vagaries of history – of geography, ecology, and chance – different for different species. The only thing universal in birdsong could be that it is under selection pressure for breeding success. After birdsong first evolved, different species went their separate ways, each influenced by selection pressures from its own environmental stresses, conditions, and events. Today we see traits that to some extent are ghosts of situations that each species encountered in the past. A song trait that may have been adaptive under specific circumstances long ago may be only a historic artifact today. Such non-functional traits would be subject to ongoing random genetic drift, masking further any meaning, and making any search for explanation futile.

And so, each species is a composite of its own history, of the way it used song to contend with various ecological circumstances over hundreds of thousands, even millions of years. Why should we expect logical explanations for everything we see today? Now, much birdsong may just exist – acoustic variability and richness without reason.

But maybe such an explanation is a cop-out. Future research will tell.

Behind all this variation in birdsong is a physiological ability to produce the most complex sounds in nature. Variation can happen in both pitch and volume at speeds that greatly exceed detection by the human ear. We simply cannot engineer any equivalent. A Swainson's thrush, for example, sings a song that includes about forty notes per second. The best we can do with the human voice is about four, and with all ten fingers on a keyboard maybe twenty.

Experimental evidence shows that birds can and do receive this fast-paced messaging.

Birds evolved from a branch of dinosaurs 150 million or so years ago.[23] Nobody knows if their immediate ancestors ever sang, because singing primarily involves soft tissue that has left no fossil record. Before birds, some amphibians may have produced musical notes, but few insects are musical, and early reptiles and mammals undoubtedly relied on vision or scent to repel rivals and attract mates. The world must have been a silent place; no music enlivened the wind as it sighed through Mesozoic tree ferns and giant horsetails.

The mechanism of birdsong – that is, its physical apparatus – is unique in the animal kingdom, with no analogue either in mammals or in human instruments. Nonetheless, research into this mechanism has extended knowledge about other aspects of basic animal functioning, including topics that are very relevant to humans, such as biological clocks and imitative learning.

The songbird's vocal instrument is the syrinx. It consists of a tympanic membrane that extends into the wall of the bronchial tube and vibrates, breaking up the outward passage of air into waves. The frequency, or number of waves per second, sets the pitch of the sound. The size of the wave sets its volume. Up to a dozen tiny muscles adjust the tension on the tympanic membrane to alter both pitch and volume, doing so at millisecond speeds.[24] Controlling these muscles are nerve signals from a group of sensitive neurons that form a "song centre" in the forebrain. Specialized song-learning neurons are included in the oscines but are generally absent in suboscines.[25] This song centre is pronounced only in males.

Making the whole apparatus even more remarkable is that it is duplicated, so that a bird can sing its own duet. Songbirds have a split syrinx, with one fully functional voice box in each of the two bronchi coming from the lungs. One of the best duet singers with itself is the veery, and a sonogram of its song shows its two streams of simultaneous music, one at a slightly higher pitch than the other. That feature gives its song a hollow, singing-down-a-drainpipe quality.

Clearly, an apparatus so complex, rivalling the complexity of the vertebrate eye, has been worked on and improved over the millions of years of songbird evolution. Yet the marvel of anatomy is only the beginning. It is one thing to possess a good musical instrument but another to be able to play it. Doing so involves not only the song centre in the brain but a whole set of neural and hormonal pathways.

The song centre increases in volume during the breeding season, and then shrinks. It is controlled by hormones from the reproductive organs. These hormones, in turn, are under the control of a biological clock that is regulated by the length of con-tinuous daylight.[26]

Most animals – from fruit flies to humans – and even many plants have a biological clock. It adjusts a host of metabolic and physio-logical activities. Jet lag reminds us of its existence. Hunger, sleep, and digestion all are regulated by a twenty-four-hour clock. Where is the clock in animals? How is it regulated? How does it work?

Because light is involved, so, obviously, is the eye. Not only is the eye sensitive to light, however, so is the pineal organ, a small complex of cells located in different places depending on the organism. In birds it is found near the surface of the brain. Its light sensitivity was demonstrated by blind birds that responded to light after their head feathers were removed. In humans, however, the pineal is buried deep in the brain where light never falls.

In 1970, in a masterpiece of scientific sleuth-work built on a platform of wrong guesses and negative results, came the discovery of the biological clock, made simultaneously by biologists in California and Pennsylvania. The clock consists of a small cluster of cells in the front part of the hypothalamus called the suprachiasmatic nucleus (SCN). The hypothalamus had attracted attention for some time, as it is the primary nerve switchbox in the brain, responsible for sending nerve impulses to correct places.

This discovery raised the question of just how the SCN works. Over several decades, an answer has emerged. Nerve impulses travel from the retina of the eye to the SCN, which responds by firing off impulses along a pathway to the pineal gland. The SCN also produces hormones that are dumped into the bloodstream (as are all hormones) and by that means also reach the receptive pineal gland. So, the early research on the pineal in birds was at least partially correct – it provides part of the clock mechanism.

The pineal responds to these signals by producing the hormone melatonin – yes, the same melatonin that is available in pharmacies and is prescribed to help people with sleep disorders. Melatonin is dumped into the bloodstream and, among other physiological effects, it reaches the song centre in the brain.[27] The song centre has direct nerve connections to the muscles of the syrinx – and the bird hops up on a branch and sings. All this is very mechanistic, and the bird sings not at all because it is "happy." In fact, it could be very annoyed about a rival, or a recalcitrant female, or maybe life in general.

Besides helping to elucidate the working of the biological clock, studying this song mechanism has had other payoffs. It has been a research vehicle for understanding the processes of learning, something that interests humans because we carry it further than any other species. One form of learning involves imitation, employed for vocalizations not only by birds but dolphins, whales, possibly social carnivores and primates, and of course us. Imitative learning rests on a genetic predisposition to learn, and in the case of birds, to sing.[28] When a young oscine bird hears its father sing, its genetically programmed predisposition is to repeat what it hears. This environmental programming layers on top of, and puts controls on, the fledgling's ability to produce a variety of unregulated sounds. If a young oscine bird is deprived of hearing its father sing, it will produce a variety of sounds but not a correct song. As it learns, it disposes of the sounds it does not hear and eventually produces an appropriate song.[29] Much the same process happens in human infants learning to talk.

The genetic predisposition for imitative learning appears to be based in part on a gradual, evolutionary transformation from learned information into selected genes in a process called "genetic assimilation." This may be a key cognitive process, responsible for much instinctive behaviour in animals – there is still much for us to learn. Genetic assimilation happens when those individuals in a population that learn a specific behaviour the fastest and most accurately have some selective advantage. Maybe they are also the most alert and survive predators the best, or maybe they are more vigorous and therefore more attractive to females. For whatever reason, if a selective advantage is involved, previously hidden differences among individuals are exposed, and gradually the ability in the population to learn the activity improves. After many generations, some individuals may respond so quickly that the behaviour – here the song – becomes instinctive or innate.[30] In suboscine birds, whose songs are instinctive, perhaps this process has proceeded to such an advanced level that song learning need not occur. In the oscines, however, learning is still required.

Most genetic assimilation in animals does not result in complete instinctiveness but will be partial, leaving some room for learning. Nature and nurture act together, as has been known for some time, but now we understand the linkage better. We can indeed inherit a propensity for a certain skill, be it playing the piano or playing hockey, if that skill involves motor movements or mental processes that have had survival value in the past. But the right situation for learning must supplement that latent inherent talent.

Imitative learning and its codification into instinctive behaviour sound very much like Lamarck's inheritance of acquired characteristics. Once again, he may have been partially right. Genetic assimilation involves the acquisition of a trait from the environment through learning followed by Darwinian selection for that trait.[31] Its importance may be great. Instinct, or some combination of genetic disposition and learning, may guide animals to act in many adaptive ways. For example, the instinctive or partially instinctive identification of an appropriate habitat to live in, which most species exhibit, is of obvious critical importance to their well-being.

On a spring morning in 2002, we slipped the bow of the canoe into the black waters of South River. Silver wavelets, illuminated by our flashlights, spread out from the bank, vanishing into the river's darkness. Placing the packsack with our recording equipment between the thwarts, we took our places and shoved off.

South River, in central Ontario, sidles off the Algonquin Dome, pouring its tannin-laden waters northward to Lake Nipissing, the French River, and Georgian Bay. The river has watched human history parade by – the birchbark canoes of the Ojibwa, the log rafts of early loggers, the cofferdams and mill races of first settlers. Now it goes its own way largely ignored and unimpeded, too small for hydro dams, too convoluted for power boats, too many rapids for good canoeing.

The alders lining its bank shouldered out of the darkness as the first hint of day touched up the eastern sky. The current edged us along. An ovenbird sang; something had disturbed its sleep. Ovenbirds rest uneasily, occasionally calling in the middle of the night. Mary reached behind her for the microphone and propped the handle of the parabola against the bow. I switched on the recorder. The canoe drifted along.

A whitethroat sang, the voice of the northern woods, a distillation of wilderness, its crystal-clear vibrant notes slicing up the silence. We ruddered the canoe towards the far shore so the parabola would be square on when the bird sang again.

Gradually, like a developing photograph, the river ahead took shape. Only three or four canoe widths wide, it writhed and twisted through the alders. First we could only make out the limbs of big white pines along the shore, then the feathery etchings of tamaracks, finally the delicate detail of hanging birch catkins.

Deep in the forest, a hermit thrush sang – melodious, flute-like, ethereal. I warped the canoe in the thrush's direction; this bird was one of our research subjects. Another whitethroat sang, and an olive-sided flycatcher. A beaver slapped its tail and dove. We did not talk, just silently drifted along. Now and then, the current murmured as it slipped over a log or gurgled around the near side of a curve.

Almost imperceptibly the daylight and birdsong intensified. More whitethroat song; this is one of the most common birds of the Canadian Shield. More hermit thrushes, and then the haunting, hollow, cascading notes of a veery. In counterpoint, a Swainson's thrush made its entrance with its ascending, liquid notes. At first he was distant, but as we drifted closer, his song made the recorder needle jump. In pine country such as this, thrush song pours out each morning and evening like liquid gold.

A ruffed grouse drummed from just behind the alder fringe. Swamp sparrows began jamming from both sides of the river, and yellow-rumped warblers, chestnut-sided warblers, magnolia warblers, scarlet tanagers, rose-breasted grosbeaks . . . The level of sound swelled into a triumphant symphony. We let the music roll over us until the noise of rapids ahead warned us to head for shore.

And then, while the wavelets rocked our canoe, we talked about the morning's performance. "Great." "But do you think it was as good as you expected?" "Well no, not really, not the way I remember it." We both had the uneasy feeling that, compared to the days of our youth, the dawn chorus had diminished. The thought was chilling. Was the dawn chorus no longer the equivalent of a full philharmonic orchestra but only a small-town band?

That, unfortunately, is now one of our research questions, although answering it may take several decades. Long time spans are usually necessary for ecological monitoring – maybe. If birdsong has diminished, then it would be consistent with mounting evidence from banding stations and other sources that the populations of many songbirds of North America, particularly of the eastern forests and grasslands, are decreasing.[32] Moreover, rapid climate change is predicted to throw migratory birds out of synchrony with conditions such as food supply at critical stopover places, possibly ending the era of mass migration that has been a global feature for millennia.[33]

So, ever since that morning, each spring we repeat that float trip several times to strike an average, because the level of song on any one day is partially linked to morning temperatures. We download these recordings into a sound software program that allows us to count the number and duration of gaps when no song occurs.

Memory is not good enough. That is why environmental monitoring is so important – to provide benchmark data. Without tracking change, we might wake up some morning and realize the world has irrevocably changed. Will the chorus of birdsong, one of nature's masterpieces, have slipped away?

Right now, however, it is satisfying to realize that a dawn chorus is still happening somewhere in the world almost all the time, 365 days a year, as it has for tens of million of years. In North America the chorus sweeps east to west over a four-hour period with the moving morning, then crosses the Aleutian archipelago and, with an almost imperceptible break, makes landfall again in Siberia and Japan, before sweeping across Asia and Europe, breaking again only briefly as it crosses the Atlantic to start over again the next day. That happens progressively in the northern hemisphere from February to the end of June as spring moves north, and in the southern hemisphere from August to December as spring moves south, with the extended breeding season for birds in the tropics filling in the missing months. As John Muir wrote, "Eternal sunrise, eternal sunset, eternal dawn and gloaming, on seas and continents and islands, each in its turn, as the round earth rolls"[34] – somewhere a dawn chorus of birdsong is always happening. May it never fizzle out.

CHAPTER FOUR

THE PERFECT MOOSE

Out of the ecological and evolutionary forge have come species of seem-ingly overwhelming perfection, complexity, and beauty, each one shaped and moulded as an integral, inseparable part of its environment. So many intricate devices, so many clever designs, and all bestowed by that cunning, cruel, magnanimous force – natural selection. Across thousands of generations for young species, millions for old ones, selection has re-spun every adaptation until it has reached greater and greater perfection – tried, tested, and true. But . . .

HOW IMPROBABLE IS IT that a colossal, one-tonne animal, decked out with two-metre-wide antlers, should have evolved in dense forest where it has to escape often enough from an accomplished and swift predator like the wolf, play host to a variety of parasitic worms and biting flies, and, to top it off, be a "resource" for exploitation by humans? As added handicap, it is energy-inefficient in deep snow and finds unpalatable the spruce trees

[76]

that dominate the boreal and montane forests where it lives.

How improbable? Just enough to become a huge evolutionary success, to emerge as one of few large mammals to survive beyond the Pleistocene Epoch, then rise to prominence as one of the north's greatest forms of animal biomass.

Tens of thousands of years ago, moose in Siberia shared northern forests with other megafauna the likes of mammoth, mastodon, woolly rhino, Irish elk, cave bear, short-faced bear, and stag-moose. All these compatriots, however, eventually died off, largely victims of an unprecedented two-legged predator – us.[1] Some 9,500 years ago, moose migrated across the Bering land bridge and became opportunists in a new land, exploiting the vacant niche previously occupied by the slightly larger stag-moose (*Cervales scotti*) that had vanished a thousand years earlier. Specific niches persist, but species exploiting them change, passing the baton of life from one runner to the next.

When moose arrived in North America, the sabre-toothed cats they had once known in Eurasia had departed from both continents.[2] Extinct, too, was the North American lion. Still present and eager to greet them, however, were gray and red wolves of the New World, and farther south, large dire wolves in the twilight of their existence[3] and, of course, recently arrived humans.

We have studied moose for many years. Moose have knocked down our tent, chased us up a tree, and almost upset our canoe. We have collected their bones at wolf kills, analyzed their marrow fat, indexed their densities, described their habitats, and spent hours in clouds of blackflies running browse and pellet surveys to determine their previous winter's food and cover needs.

For us, moose have emerged as a symbol of survival against the odds, success beyond the expected, and a species suitable to address the question: To what extent is there "perfection" in nature? Even before Darwin, the oft-discredited Jean-Baptiste Lamarck had put forth the notion that evolution is progressive, that things get more complex and perfect.[4] Darwin himself was ambiguous about perfection, referring to it on occasion as a result of selection.[5] In 1926, at the famous Scopes Trial in Tennessee, where the theory of evolution itself was put on trial, Professor Horatio Hackett, zoologist at the University of Chicago, speaking as a witness for the accused (arguing that evolution is a fact), explained that "the sequence of these stages [in the evolution of the horse] in geological time exactly fits in with the theory that each one has been derived from the one next below it by more perfect adaptation to the conditions of life."[6]

The fact that evolution through natural selection has trended towards greater complexity is well accepted. But the question remains: Can natural selection craft perfection? It seems almost axiomatic that if natural selection weeds out the poorly adapted, then the remainder are more perfect, and the species alive today represent current pinnacles of perfection. Is that interpretation valid?

Our timing was impeccable – and lucky. We emerged from an alder-choked trail to the rim of a small pond at 5:58 p.m., only moments before a majestic bull moose stepped out, across the pond, from the same alder-choked trail. Had we arrived seconds earlier, we would have scanned the pond, concluded nobody was home, and walked into full view. Then, we would have heard the hooves of a startled moose pounding off in the other direction.

We had approached the pond carefully; moose tracks on the trail were "steaming fresh." The instant we arrived, moose antlers clacked against alders, a surprisingly loud sound that can carry a kilometre or more in still air. Then antlers, head, and shoulders appeared. The big bull looked around, saw nothing disturbing, and strode regally through golden sedges to the water's edge. He paused again, as if posing for a picture, then waded in.

Despite appearances, moose have a lot of things going for them, each trait helpful but in itself not sufficient for success. These traits illustrate the metaphor of an "adaptive landscape" first advanced in 1932[7] and still useful today. Imagine a mountain range with peaks and valleys. The peaks represent different traits that lead to reproductive success; the valleys represent more limited success. Each individual moose occupies a different position on each adaptive peak. If it is big-bodied it is near the top of one adaptive peak; if it possesses keen senses it is near the top of another; if it has large antlers it is high on another. Other peaks represent strength and stamina, or digestive efficiency, or a good sideways kick, or . . .

For any desirable trait, natural selection pushes individual moose uphill. While this upward push can be interpreted as movement towards perfection, there is a complication that becomes evident when considering the various traits of moose separately.

Bigness, for example. Bigness has repeatedly earned a place in the sun, not only in the dinosaurs and the mammal-like reptiles, but also in successive waves of large mammals over the past 65 million years. Many huge species have crossed the world stage, the likes of Indricotheres – ancient rhinoceroses weighing in at fifteen to twenty tonnes, largest land mammal ever to live – and Entelodonts, primitive one-tonne pigs. Bigness means strength, stamina, larger, less vulnerable calves, big brain, big energy-storing gut, and, except where adapted to stand and defend, greater speed.

The principal driving force for all these traits was, and is, to beat the predators.

How bigness arose in response to predation is easy to envision, the product of the famous "Red Queen Effect." In Lewis Carroll's *Alice in Wonderland*, the Red Queen explains, "Now here, you see, it takes all the running you can do, to keep in the same place." This quotation was picked up as an analogy to illustrate one aspect of evolution: prey escape from the predators better if they are bigger and faster, which puts pressure on the predators to get bigger and faster, which, in turn, puts pressure on the prey to get even bigger and faster . . . [8] In dinosaur days, big predators such as *Tyrannosaurus rex* drove the evolution of big prey like Torosaurs,[9] and vice versa. In the Oligocene, 35 to 23 million years ago, big predators such as dog-like Hyaenodons and bear-dogs drove the evolution of big prey like Indricotheres, and vice versa. In the Pleistocene, big predators such as sabre-toothed cats, dire wolves, and short-faced bears drove the evolution of the big, fast prey like Irish elk, stag-moose, wapiti, caribou, and of course moose, and vice versa.

Large predators drove the evolution of slow, cumbersome, giant herbivores, too: mammoth, mastodon, cave bear, ground sloth. They all outstripped their predators in the Red Queen race, winning with hugeness that for some reason the predators were unable to match. All the slow and cumbersome beasts, however, eventually dropped the baton, too. They were not ready for weapon-wielding humans. Huge body size without speed disappeared from the northern hemisphere, although it still exists in tropical-living elephants, rhinos, and hippos. As the co-discoverer of natural selection, Alfred Russell Wallace, put it, we live in "a zoologically impoverished world, from which all the hugest, and fiercest, and strangest forms have recently disappeared."[10] Yet moose managed to survive.

The big bull waded towards us, repeatedly pulling up mouthfuls of aquatic vegetation. He was still unaware of us. Usually we see the rear-end view of a moose running down a logging road in the headlights of our truck, or crashing off a portage trail, or a distant view of one wading in a pond. But what made this moose encounter perfect was the scene. Behind him, flaming above the dark shoreline conifers, rose a hill of autumn splendour – oranges and reds of maples, yellows of birches. Farther back rolled ranks of other hills, their brightness mellowing with distance. Reflected brilliance cut slashes in the ink-black pond water; the rich ochre of sedges mirrored the shore.

Tracing the evolution of big body size from the dim past in moose, or in any species, raises difficult interpretive problems. The evolutionary history of New World deer (which includes moose) is bedevilled by a scarcity of fossils. The best illustration of increasing body size is found in the ancestry of the horse. As the Red Queen Effect predicts, early progenitors of today's horse were small, dog-sized creatures in the Eocene forests of 50 million years ago, which ate soft fruits and leaves. Their descendants included an array of variously sized three-toed animals with stronger teeth that allowed them to eat grasses on the Miocene plains 15 million years ago. Among these arose one of increasing size that emerged 4 million years ago as today's genus *Equus*.[11]

The history of the horse is complex, however, involving two processes of speciation: incremental change and splitting. To understand incremental change, imagine a selection pressure for increased body size working on a beaver population averaging roughly 18 kilograms (40 pounds) per animal, as they weigh today. The hypothetical selection pressure is slight, resulting in a mere increase of 0.0045 kilograms (1 pound) per generation. Even that small amount would reflect rapid change in evolutionary terms. If that percentage increase (0.00025) were to happen each year, the beaver population would reach the size of the giant beavers of the Pleistocene, which were roughly the size of black bears, in approximately 900,000 years. If fossils representing every 300,000 years were found, which would be an impressively frequent occurrence, a paleontologist would likely classify the three individuals as three different species, especially because increasing size is inevitably accompanied by skeletal adjustments that exaggerate differences. Yet with a more complete set of fossils, collected, say, every 10,000 years, each percentage change in body size during these shorter intervals would be only between 1 and 2.5 percent, leading to an interpretation that they were all the same species undergoing incremental change. No guidelines exist to specify when species subjected to incremental change warrant the label "new species." Any effort to define this would necessarily be arbitrary and therefore invalid.

Complicating interpretations of speciation further is the entirely different process of splitting, in which a lineage branches in two. For splitting to occur requires isolation of part of a population, followed by genetic change.[12] Unlike incremental change, this process results in an increase in the number of species because the original one persists.[13] Evidence of splitting is a simultaneous fossil record of both the progenitor and the new species.

No small, beaver-sized moose has been found in the fossil record, contrary to expectations from the Red Queen Effect. About 2 million years ago, an animal identifiable as a moose, and placed in the same genus as today's moose, arose from a lineage of moderately large progenitors that had obviously split several times in the past. This early moose was already as large as a yearling Alaskan moose today.[14] The Red Queen Effect must have worked to a considerable degree on moose ancestors. But, as with the complex history of the horse, both processes of speciation – incremental change and splitting – seem to have been involved in some convoluted way, and we are left today with the large descendent of what was once a small ancestor.

At any second we expected the moose in the pond to notice us, but with a darkening forest behind us, we presented no silhouette. His newly velvet-free antlers swept upward, and as he slowly lowered his head, the antlers formed a mirror image in the water. Belly deep by now, he immersed his mouth, nostrils, sometimes even his eyes so that only the skull-bulge at the top of his head was visible. We timed his periods of submergence – up to thirty-seven seconds. Occasionally he would exhale underwater, blowing clearly audible bubbles that frothed around his snout. Once, he shifted to high alert, looking down the far shore, but concluded that it was a false alarm and resumed feeding. Alertness had so far kept him out of wolf stomachs.

Not only evolutionary history but characteristics of the immediate environment are necessary for a species to climb an adaptive peak – that is, to actually use its genetic potential. Moose, like other large animals, are helped up the peak of big body size by nutrition. Moose are habitat opportunists, seeking out disturbance landscapes where new plant growth is most nutritious. River flood plains, for example, grow willows with high levels of protein. Fresh avalanche slopes with new deciduous growth, early-succession forests after fire, beaver ponds with aquatic vegetation, and even logged-over areas with abundant regrowth – moose benefit from them all.

Ice age landscapes provided, and still provide, many of these conditions, where wind-blown sediments are scoured out of rock and dispersed across the land as highly fertile soil. Glaciers, too, revitalize nutrient cycles by depositing fresh soil. The soils of tropical landscapes, in contrast, are leached by years of heavy rainfall, making nutrients less available. Various tropical herbivores are under selection pressure from predators to grow large, too, but, except for a few notable species like elephants, inadequate nutrition may impose limits.

Nutrition may be an underlying reason for the size differences among the four subspecies of moose in North America today.[15] Capitalizing on the good nutrition that characterizes more recent ice age environments, an Alaskan bull may weigh over 450 kilograms (1,000 pounds). The soils of such environments typically are enriched by wind-blown deposits from newly pulverized rock, driving plant productivity to the limits that cold temperatures will allow. In contrast, an Algonquin bull weighs 90 kilograms (200 pounds) less. If these differences are nutritional, as suggested, then some of these subspecies should be reclassified, from a genetics standpoint, as just one.

The big bull was still unaware of us, intent on his evening meal. Rarely in a close encounter do we think about the complexities of moose evolution. The immediate sensory impressions are too strong. We wonder, instead, "Is this his normal evening ritual? Was he resting under cover all day, waiting for dusk to feed? Has he thwarted any recent wolf attacks?" Only later, back at the cabin or sitting around a campfire, do

*we think about his marvellous adaptive fitness – what we know, or think
we know, and what we don't.*

Large antlers form another adaptive peak, increasing the likelihood of a moose leaving more offspring. Unlike body size, large
antlers are not the product of Red Queen selection. They would
have been had they evolved as a defence against predators. Instead,
their primary function is display, both to other males, for dominance, and to females, who, as in most species, are the sex that
selects the mate. Directed towards either sex, antlers help individual moose perpetuate their genes. As such, antlers represent
sexual selection as Charles Darwin surmised.[16]

Like large body size, the evolutionary history of antlers in
hoofed mammals took place in nutrient-rich, former ice age landscapes. If primates had evolved in such landscapes instead of the
tropics, we might be adorned with antlers, too. Because of our
tropical origins, we have no such fantastic display organs.

Female moose are attracted to large antlers for what they represent: health, vigour, and competence at foraging – desired traits
for her offspring. Those females that pick mates with these traits
will simply have more viable offspring than those that do not.
Similarly, deformed or asymmetrical antlers may mean a parasitized, diseased, injured, or stressed male, as any smart female
moose inherently recognizes.

Antlers also are used for sparring, a form of ritualized and non-
lethal aggression between rival males. Again, large antlers are
advantageous, acting as shields to catch and hold the opponent's
attack. They serve, as well, as a signal of social rank that may
thwart sparring even before it begins. And they are used for rat-
tling the shrubs to announce the male's presence to females.

By now the moose was only about fifty metres away and still had not detected us. No breeze carried our scent to him. If moose were unable to go about their business peacefully, the stress of living in an eternal combat zone with their predators would be unimaginable. Maybe a perfect moose in his prime, like this one, has a self-confidence about his ability to ward off attackers.

Is there such a thing as a perfect moose? Is there perfection in any species? Can a species find itself simultaneously perched on top of all the adaptive peaks? Herein lies a biological constraint. A species cannot maximize its adaptive traits for everything. Moving to the top of an adaptive peak for one trait must create some slippage downward on other peaks for other traits. The result is a sort of mid-slope average. This situation is like water in a set of interconnected, vertical tubes. Draw up the level in one tube and it will fall in the others. The levels may not all be the same within an individual, because what gets drawn up highest is normally a response to the greatest selection pressures at the expense of lesser responses to others. A species must be looked upon as the best possible combination of traits.

For example, there are costs to bigness. Quick acceleration is lost with size. Once, we radio-tracked a pack of wolves in deep snow to a carcass of a big bull moose. As we approached, the wolves faded back, as usual, and stayed a few hundred metres away while we examined the kill. Tracks showed that the wolves had fanned out and simply charged the unsuspecting moose, who had managed to take only a few steps before the pack was on him. He'd had neither the sudden speed nor the agility to respond quickly. Often, very few seconds and very few metres determine the outcome of attacks, as we have seen repeatedly in encounters between Yellowstone elk and wolves. An upper limit on moose size, or its height on that adaptive peak, may be limited by the need to have some height on another adaptive peak, particularly an ability to accelerate when attacked.

So obvious is the necessity of compromise among traits that it has its own name: "environmentally stable strategy" (ESS). A species attains a level of fitness with its environment that represents the best stable expression of all its adaptive traits.[17] Dean of evolutionary biologists Ernest Mayr put it this way: "The morphology of every organism reveals to what degree it is the result of a compromise."[18]

A trio of gray jays sailed over the pond, softly calling, but the moose was absorbed in his evening meal and paid them no attention. A raven flew by, heading for its night roost in some distant hemlock. A few black-capped chickadees probed the balsam firs. An evening breeze stirred the flaming leaves and eddied off in the alders. Still the moose was intent on rooting up water arum and other aquatic plants.

Antlers entail compromise, too. Antlers represent a huge annual energy investment, because, unlike the horns of bovidae (cattle, muskoxen, goats, sheep), they are shed annually. Even greater than the energy cost of their production is the energy cost and inconvenience of carrying them around. The antlers of a typical Ontario moose weigh 14 kilograms (30 pounds), an Alaskan moose 23 kilograms (50 pounds). What could be worse, when down in the alders, than having to manoeuvre with such a handicap? But antlers illustrate a conflict between maximizing natural and sexual selection, which at times may be at serious odds. Moose cannot avoid the dense alder thickets that rim most beaver ponds and streams in the boreal forest and cover logged-over sites in mountainous terrain. Nor can they avoid the dense willows on the subalpine slopes and river flats of the north. Both alders and willows are preferred foods. Yet a big bull with close to a two-metre spread of antlers, if attacked in such a place, is clearly in trouble.

Any antler-induced slowness in running from predators, however, may be offset by a very different survival strategy – standing at bay – and so important is that trait that it forms another adaptive peak. Moose that stand and face a pack of wolves most commonly survive. Not always, or all moose would defend themselves exclusively in that way. However, moose have evolved a lethal kick, not only backwards with the hind legs but to the sides as well. We have seen evidence of these kicks during autopsies on wolves, most commonly as broken and re-fused or partially re-fused ribs, but on two occasions as fractured and re-fused leg (metatarsal) bones. Occasionally, wolves are killed by moose. We found one wolf crumpled in the snow with a cracked skull and two broken ribs, amid a mass of moose tracks. Where wolves have attacked a moose, typically, the forest is torn up and blood is spattered around, evidence of a standoff rather than a clean kill. Similarly, we have seen bull elk in Yellowstone successfully defend themselves against a pack of wolves, sending them flying with a well placed kick.

Evolution has failed to forge perfection in another way, one that presents a conundrum. All living things are mortal. Each one wears out, breaks down, something anatomical or physiological goes wrong – that is, if its imperfection hasn't meant that earlier it was the loser in some battle against a pathogen or a predator, or it hasn't fallen over a cliff. Once, while walking through a spruce forest at the base of a cliff, we glanced up and were startled to see a hanging, partially decayed moose. One hind leg was pinioned by a tree root that looped out and back to the clifftop. The previous winter's snow had hidden this natural trap, and the unfortunate moose, not recognizing the edge of the cliff, had taken one last step and pitched forward into mid-air.

Why, after investing energy in growth and development to reach some optimum adaptive and reproductive state, do organisms deteriorate and die? Natural selection should favour individuals that live longer and leave more offspring. Instead, as Thomas Hobbes said in the seventeenth century, life is "nasty, brutish, and short."[19]

At the heart of this conundrum seems to be the sacrifice of the individual for the welfare of the population. Death makes way for population turnover. Each generation re-shuffles the genetic deck, and this, in turn, allows populations to meet environmental challenges and change. Dramatic evidence is provided today by many pathogens and insect pests that beat our best efforts to eradicate them. Without death and subsequent replacement, entire species adapted to earlier conditions would be doomed. Besides, without death, reproduction would quickly flood any ecosystem to the point of resource depletion and collapse.

This explanation of death – to provide room and resources for the next generation – was proposed by August Weismann in the 1880s. But many biologists today would object to this reasoning. They would argue that natural selection works on individuals and cannot foresee any consequences for populations or species, because no mechanism has been identified to provide that extension of benefits (an idea explored further in Chapter 13). The welfare of populations and species can be no more than a by-product of individual welfare. Normally, individual welfare translates into population and species welfare, but not always. There may be a disconnect.

Illustrating one form of disconnect – that of individual gain at a cost to the population – was an annual winter migration of white-tailed deer that moved from deep snows in Algonquin Park and concentrated tenfold in a smaller area of lighter snow to the east. This movement served to minimize energy loss over the winter and improve nutrition for individual deer. But many of the wolf packs also left their home ranges and followed the deer. Here was an ideal set-up for the predators, with less search time required to find prey.[20] The deer population suffered greater mortality, even though individual deer, partially protected as members of herds and with better nutrition, reduced their own susceptibility. A similar concentration worked against a caribou population that we studied in central Labrador (see Chapter 8).

Death reflects the reverse disconnect – individual loss but population gain. Because natural selection simply cannot direct such a thing, how can it happen? Considerable research has gone and continues to go into answering this question because of its relevance to human gerontology. Two explanations were put forward almost fifty years ago, and they are still relevant today because of increasing experimental evidence.[21] The two explanations are variants of each other, both resting on the obvious fact that once an organism is too old to reproduce, natural selection is irrelevant. Nothing that happens later will influence the number of its offspring, except for the rare situation in which grandparents play a significant role in raising grandchildren. The grandfather in one wolf pack we studied in Algonquin was the most frequent babysitter when the other adults went hunting, although how vital his role was to the welfare of the pups was uncertain.

One explanation for death is that late-acting genes can reduce an older individual's fitness, and eventually prove fatal. They can accumulate in a population and be passed from generation to generation without being purged, as would happen if these genes were operating on reproductive individuals. This explanation was put forward by Nobel Prize winner Peter Medawar. It was followed by a second explanation, a "pay later" theory by George Williams, based on the biological fact that a particular gene may have an effect on several traits. He proposed that specific genes may help maximize vigour and reproductive output in young individuals, but then turn around later and contribute to the individual's demise. As experimenters worked on that idea they found a relationship between fecundity and longevity, as if the allocation of energy to reproductive output resulted in a later cost of cellular breakdown.

In recent years, specific "death genes" have been discovered. Their removal has increased longevity in laboratory populations of fruit flies, guppies, and mice – 30 percent in the latter. Intensifying interest in these findings are new techniques in genetic engineering and transgenics that magnify the possible application to humans.

Today, we are still left with Weismann's observation that death of individuals is essential for populations and species to adapt, and individual death, according to all three theories, results from an inevitable biochemical breakdown that even natural selection, working as it does to maximize individual reproductive success, cannot correct. Ernest Mayr, who himself denied the lethal genes of old age until he was 100, wrote in his second last book, published when he was 96, that "selection cannot eliminate the genetic propensities or diseases of old age."[22]

Current research on this topic may be biased by thinking too much about our own species. Unlike most species, we have a prolonged post-reproductive life. Other than us, natural selection has had a limited larder of post-reproductive-age animals to work on. Most wild animals die before their maximum biological longevity, as shown by comparisons between wild and captive animals.[23] Moose are typical. Only once have we encountered what was likely a post-reproductive moose. Even this big bull did not die of old age, but when Algonquin wolves killed him, senescence had made his demise inevitable. He was a Methuselah, aged by tooth sectioning at an ancient fifteen years. His bone marrow contained almost no fat; it had all been metabolized in a physiological effort to commandeer enough calories for tired legs when deep snow was sapping his strength.

In contrast, the moose mirrored in the pond that evening was in his prime, likely five or six years old. Even he was lucky to have lived that long, given the host of environmental threats he faced. We have collected the bones of less lucky moose that were dismembered and scattered all over the snow. As well, we have discovered frozen moose carcasses down in the spruces being chewed on by wolves, where the snow all around was littered with ticks. The winter tick, picked up innocently by moose brushing through vegetation in the summer, is a languid killer, causing hair loss and hypothermia.[24] In the evolutionary race between tick and moose, ticks at the moment are ahead. Maybe some day, moose will develop some physiological resistance, but such adjustments take time and luck. Another moose killer, moose staggers disease, is caused by a tiny nematode called a brainworm that burrows up the spinal cord and into the brain. White-tailed deer, with a longer evolutionary exposure to the roundworm – they both evolved in

North America – accommodate its movements without neural damage. Moose don't. Moose first encountered the roundworm when they crossed into North America, only recently in evolutionary terms, and they have yet to evolve any resistance.[25] Moose, like every other species, have plenty of environmental guns aimed at them, all eager to transform them at the earliest possible age into carrion.

Early death in the wild benefits even more than the victim's own population. Because of food web connections – predators, parasites, decomposers – early death helps support whole communities. By promoting rapid population turnover, its effects reverberate through ecosystems, increasing production, speeding up nutrient cycles, and contributing to biodiversity. But natural selection cannot foresee these things, either. It just goes blindly on messing around with individual animals and letting the ecological chips fall where they might.

So, we are still left with an inadequate explanation for individual declining vigour, old age, and death, even recognizing its fundamental requirement for evolution. Our current understanding is still open to new discoveries.

Other deep reasons preclude perfection in nature. Ernest Mayr summed up the subject when he wrote, "Some enthusiasts claim that natural selection can do anything. This is not true."[26] He went on to cite "the limited potential of the genotype" as a major reason. Species do the best they can (so to speak) with what they have. "There are many ecological niches that mammals have been unable to occupy. . . . The kind of genes needed for an appropriate immediate response to a new selection pressure may not be present."

———

Another factor precluding perfection is that natural selection is not just "the survival of the fittest," a phrase coined not by Darwin but by Herbert Spencer. More than the fittest are alive in all populations. Natural selection weeds out the misfits, leaving, besides the fittest, a lot of mediocre individuals.[27] Any population has a majority of them. These are the vulnerable. While they do hold a bank of genetic variability that may be useful if environments change, these mediocre individuals put the populations of all species at some degree of risk.

Finally, perhaps the most basic reason for the ultimate failure of adaptations is that the rules of engagement constantly change. Environments reorganize – they always have – and with reorganization, some adaptive peaks melt away to be replaced by others. Continents drift, climates change. Adaptations are always tenuous, always subject to adjustment. For moose – after surviving the conditions that caused the demise of most Pleistocene megafauna, after meeting the challenges of natural and sexual selection and attaining "environmental stable strategies" – along came technological humans. Suddenly, large antlers become a trophy, big body size is irrelevant, the stand-and-defend strategy is lethal.

And so, what does it mean to ponder nature's imperfections, to realize that all living things are ill designed with a plethora of foibles, that life is tenuous, that evolution represents stumbling, experimental, blind testing of biological devices, peppered with luck, over vast spaces of time? The self-organization that underpins life is not infallible. It breaks down for species where environmental change outpaces genetic, epigenetic, or behaviour adjustments. Is a hidden hand directing all this? No way! Some 99 percent of all species that have ever lived are now extinct! How could a wise creator have designed so many failures? Instead, as Stephen Jay Gould says, "evolution lies exposed in the imperfections that record a history of descent." [28]

Life's imperfections underscore that we are on our own, with no grand, preordained plan of ultimate perfection, no purpose to evolution, no secure future for the species and ecosystems around us. By changing environments, as human activity invariably does, it is axiomatic that we make less "perfect" the existing adaptations of living things. Within evolutionary time scales, that may be acceptable; change has happened before. But in the past, each rapid change invariably impoverished the world's biodiversity for a time – a long time. This is the world we live in now. If we value wild living things for whatever reasons – practical, spiritual, ethical – does recognition of their vulnerabilities lead to compassion? Does compassion lead to concern? Does concern lead to our changing how we use the Earth?

The light was fading, so when the moose put his head under again, we backed away noiselessly, then edged out of sight, avoiding twigs, stepping around a wolf scat on the trail, a reminder of the moose's probable fate. However, he was in his prime. Likely he would see his hillsides blaze red for a few more years yet – or so we thought.

Two days later, in the early morning, we again hiked to the pond, hoping that the moose had a habit of feeding there. The leaves still hung in their brilliance, touched up by a bit of frost. A tracery of ice edged the lakeshores.

We sensed trouble while we were still on the trail. Fresh ATV tracks had torn up the earth and crushed the grass. Near the pond, the alders had been slashed for access. Then, about a hundred metres beyond the far shore, we found what we'd feared – a pile of moose guts.

PART 2

POPULATIONS: SLAVES TO THE RULES

With some understanding from Part 1 of the mechanisms that organize individuals and species, particularly those that drive adaptations, we can consider the next step up in the biological hierarchy: populations. All individuals are life members of a population. That membership both endows individuals with opportunities and levies constraints. Populations are ruled by rules, sometimes benevolent, sometimes callous, always indifferent. But the rules convey order.

We have searched on the tundra, in the boreal forest, and elsewhere to observe these rules first-hand and to try to understand how they operate, why they are so strictly enforced, and, finally, how they may relate to us.

CHAPTER FIVE

GIVE ME LAND, LOTS OF LAND

The density of any animal population is adjusted by its own members through the way they claim and use land – that is, their spacing behaviour. A plethora of spacing systems are found in nature, each one embedded in the genetic psyche of the species, as much a part of them as tooth and claw. Whatever the system, it provides the population glue. It organizes and distributes its constituency across the land.

By what alchemy does it operate?

THE CASCADE VALLEY SLICES north-south through the Front Ranges of the Rocky Mountains, with the roiling Cascade River and open montane forests on the valley floor. The valley is wilderness, protected in Canada's flagship national park, Banff. Yet it is staked, end to end, in a mosaic of land claims. Invisible boundaries enclose various parcels that vary in size from less than a hectare to more than a hundred square kilometres.

The whole valley, from floor to peaks, is carved up in this way.

With each claim comes ownership and the right to exploit resources as the owner sees fit, free of rules and regulations, or almost so. Nobody oversees these activities. They are accompanied by intense rivalries, threats, aggression, sometimes even overt fighting. The peacefulness of the scene is an illusion.

Almost every vertebrate species, and many invertebrates, claim and defend land. Their action is instinctive, the product of evolutionary pressure wrought by time: to define a home range, claim a territory, or delineate a social space. Doing so is vital to survival and success, and the boundaries, demarcated by scent and song and physical presence, are as real to the species involved as condo walls or the hedge around a backyard. Individuals must lay claim to what they need to survive and produce young. What could be more fundamental to life? It drives competition and conflict like no other biological force. Books have been written on spacing demands, but we need look no further than our own species to see it working, from insults hurled across the backyard fence to wars between nations.

Understanding how spacing behaviour influences the density of populations is what drew us to the Cascade Valley to study a species that claims land, lots of land – the grizzly bear. We went as research assistants to help our daughter, Jenny, whose Ph.D. study was aimed at determining the resource needs, at a landscape level, that dictate the areas that female grizzlies select to live. Bear management issues exist throughout North America, especially in heavily visited national parks like Banff, because grizzlies and people coexist better if they maintain some separation. Understanding what grizzlies need can promote a peaceful coexistence.

Access through the Cascade Valley is along an old fire road, converted by time into a trail. Above the trail loom vertically stacked strips of subalpine and alpine country, and high on the barren, rocky slopes, the windswept firs are stunted. The east-facing slopes are steep, with avalanche chutes strewn with dwarf birch and grasses. Large U-shaped and tapered V-shaped valleys characterize the area, which features three gentle, large, mid-valley knolls known as Grassy 1, 2, and 3.

About twenty-five kilometres up the trail, along a hillside of lodgepole pines and scattered buffalo berry, was the transition between the home ranges of two females: Bear 28 and Bear 17, with her one cub. (The bears were assigned numbers in the order they were caught and radio-collared.) Nothing remarkable identi-fied this boundary, but many same-day telemetry locations from the air showed that each bear respected that invisible line. They came close at times, undoubtedly leaving scent to stake their claim, but appeared to avoid confrontation and would not arrive at the boundary simultaneously.

The core of Bear 28's home range lay to the north around a park warden cabin. One day, we received instructions from Jenny to backpack there, headquarter in the cabin, and monitor this bear's activity. She neglected to tell us that this was a grizzly epi-centre. Although the area had its own resident female, other bears could be around – and they were. Male grizzlies set up their own home ranges largely independent of females, and immature bears sometimes wander in search of their own land and mate. Even adult females accept trespass from other adult females, so that home ranges often overlap.

We encountered all these situations. Bears were so close that we dared not venture very far, and the signals and sightings effectively kept us cabin-bound for three days. Once, thinking the way was clear, we climbed a forested slope behind the cabin and, walking around a fir tree, came face to face with a big male grizzly grazing buffalo berries. We got out of there in a hurry. Another day, when 28's signal seemed fainter, we assembled our gear and were all set to go out and track her when a young grizzly appeared outside the front window. He sauntered up to the corral, stood on his hind legs, scratched against various posts, then lay down and rolled under the bottom rail. After wandering around inside the corral for a while, he rolled out again and disappeared behind the cabin.

Compensating for our lack of data, we monitored 28's signal from inside the cabin porch all night as well as all day. She wore a collar with a motion detector that varied the pitch of the signal when she moved. Some authors state that grizzlies are exclusively diurnal. This bear, however, also moved around most of the night, with only short sessions of inactivity lasting less than an hour.

But it was Bear 17 who taught us the most about the use of space by grizzlies. Occasionally we could watch her without difficulty out on Grassy 1, her dark body and golden face contrasting dramatically with the vegetation. She would spend most of a full day digging hedysarum roots or flipping rocks looking for ant larvae, staying within a 100-metre radius, while her cub hovered nearby. Then, suddenly, she would pick up and leave, her cub trailing along, walking at a steady grizzly gait towards the trees and out of sight. The next day her signal would be faint, or not in the vicinity at all. Sometimes an aerial fix would locate her up to 15 kilometres away.

Such erratic movements were puzzling, because theory on optimum foraging does not predict them. According to theory, an animal attempts to maximize energy intake while minimizing energy expenditure. If food comes in patches, an animal will exploit a patch until the supply drops to about the level of the surrounding environment and then begin to search for a new patch. But the grizzlies we watched, after exploiting a food patch, expended considerable energy travelling greater distances than necessary to another high-quality patch. We wondered why they would pass up good foraging and travel so far.

Biologists have debated the dominance of energy efficiency in the foraging strategies of animals, and our grizzly observations lent weight to the detractors. Bears have other needs, as do all species, which may override an unbending search for food. The built-in behaviour of bears is quite plastic – it can change and adapt – so that bears select landscape features to satisfy a range of needs For example, home ranges of female grizzlies in the Rocky Mountains are chosen consistently for rugged terrain, but especially so by bears classified as "wary," that is, bears with a predisposition to avoid humans. Female grizzlies with cubs show this tendency for rugged terrain most markedly. Security, then, is clearly on their minds, and a bear foraging in optimum habitat may sense danger from the presence of other bears, or humans, and will simply up and go.[1]

Demonstrating even more selectivity, Bear 17 and the others in Jenny's study favoured certain feeding sites, such as grassy meadows throughout summer, avalanche chutes in spring, and riparian areas adjacent to water in summer and autumn – but they sought those places within the context of rugged terrain. They clearly knew where they were relative to various features that meant something to them.

To learn the details of an environment so well, bears must spend considerable time exploring. Investigative or exploratory behaviour is a built-in drive in all mobile animals. Its adaptive advantages include not only knowing sources of food and security, but also finding physically comfortable surroundings, or "shelter-seeking," which is also a basic drive.

Bear 17 found all these requirements in an area of 152 square kilometres. It was oblong in shape and ran northwest-southeast to fit the Cascade Valley, extending to the mountaintops on both sides. She did not own this chunk of land exclusively, however. She shared the northern quarter with Bear 28, whose home range encompassed 409 square kilometres, and another portion with Bear 33, whose home range was 797 square kilometres. Other females using the Cascade Valley held home ranges of 895 and 884 square kilometres each, but these larger ranges straddled mountaintops of bare rock and ice that were rarely used.[2] The sizes of home ranges for females on the eastern slopes of the Rocky Mountains varied from as little as 152 to as much as 1,413 square kilometres.[3] In contrast, in the Khutzeymateen, British Columbia, an average adult female home range is around 53 square kilometres; areas in coastal Alaska average 522.[4]

Overall, the use of space by grizzlies fits the definition of "home range" better than "territory," the difference being that home range lacks the higher level of defence that results in exclusive use, especially in boundary areas. Possibly the core of each bear's home range is defended, resulting in a classic territory within a home range, but evidence for grizzlies of such an area-within-an-area is weak.[5] The extent of overlap between bear home ranges varies, increasing or decreasing with food availability. High-density populations exist where food is abundant and uniformly distributed, and vice versa.

These space demands of grizzlies do much to determine their population density. Their home range size is set large by resource demands and restrained by competition. They cannot be crowded – or can they?

A damp chill permeated the morning air. The sun's rays touched up the willow leaves still dripping from the morning dew. Autumn had slipped imperceptibly through the spruce forest. Willows and birches glowed golden, bearberry leaves flushed crimson, and sweetgale's pungent scent wafted on the breeze.

A golden-mantled brown bear sow leisurely herded her three half-grown cubs through the water into the river shallows and onto the trail. A sighting like this would be a highlight anytime. But these bears were not alone. Behind the family in the shallows, two juveniles were engaged in a friendly wrestle. They play-bit each other, circled, sniffed, heads low, signalling by attitude and posture their peaceful intent. Gradually they manoeuvred onto the shore and continued their friendly sparring. Standing as if in an embrace, they struggled until one bear lost its balance and collapsed onto its back, its huge paws pedalling the air.

We were in Katmai National Park, known as the "Land of Ten Thousand Smokes," a land of extremes, a land layered with the debris of cataclysmic volcanic eruptions. Katmai, at the foot of the Alaskan Panhandle, is a part of the peninsula that leads to the Aleutian Islands. It wasn't just the land that had lured us here, though. It was primarily *Ursus arctos* – the grizzly.

For years, we had observed grizzlies through spotting scopes at comfortable distances. At Katmai, however, we could observe them up close. Katmai is one of three major sanctuaries in Alaska that protect brown bears and their habitat, with an estimated 2,000 living in an area of 14,870 square kilometres. (Brown bears and grizzly bears are the same species. The name brown bear is used more commonly in coastal Alaska.)

It was early September when we took off from the little fishing village of King Salmon in a Beaver float plane for our destination, Brooks Camp, headquarters for the park. Below the grey sky, our small plane skimmed the flat, spongy tundra, eventually circled an emerald lake, descended, and taxied in.

We waded from the float plane to the water's edge, then shouldered our packs and headed down the shore, but a shaggy hump thwarted our progress. It was a mother brown bear lying with her two cubs in the middle of the beach, their coats incandescent as a shaft of light opened on them. They were barely thirty metres away, ordinarily much too close for safety. However, they appeared indifferent to our presence so we slowly worked our way into the trees next to them, and the sow let us pass.

Soon after our arrival at Brooks Camp, we discovered that brown bears roamed everywhere, using the trails and the beach by the campground as a highway to and from the Brooks River. Katmai's park biologist estimated that seventy bears were congregated along the two and a half kilometres of the river during our stay. Incredibly, the bears have developed a remarkable tolerance towards the curious, gawking, shutter-snapping humans. In Katmai, bears have the right-of-way, and humans must adjust their travel routes, times, and actions accordingly. The park wisely regulates how closely visitors can approach, ensuring the safety of both wildlife and viewer and also helping to maintain the delicate relationship among the large number of bears themselves.

Food – salmon – had caused the concentration of bears. For hours, over several days, we watched the "bear buffet," mostly from two elevated viewing platforms along the river. Some bears waded shoulder deep with heads bent and eyes submerged, searching for salmon. Others caught fish while swimming or "snorkelling," submerging their bodies and emerging eventually with their prize. Salmon die a few days after spawning, and their carcasses float downstream, eventually settling on the riverbed. Taking advantage of this situation, some adult bears ran back and forth in the shoals scooping up dead or dying fish with their paws while sows with their cubs gorged on salmon remains at the river's edge. Live salmon provide about 4,500 calories, dead ones around 1,500, but even the carrion represents a bonanza, not only for the bears but also for a host of scavengers – gulls, foxes, and ravens – that come to the river banquet.[6]

Superficially, this concentration of bears along the Brooks River seemed like a random assortment, with individuals coming and going according to their own whims, independent of all others. This was not the case. The concentration was, in fact, highly organized, the result of hierarchical behaviour, one of the classic short-range spacing systems commonly found in aggregating vertebrate populations, particularly social ones, from packs of wolves to barnyard chickens. Hierarchical behaviour conveys social order, and social order is a raw material for natural selection. It separates the biological best from the less fit and metes out success or failure.

At Brooks Camp, it soon became apparent that the bears were spacing themselves out according to size, sex, and behaviour. They were using behavioural signals that represent a language of aggression and submission. Males were at the top of the social order, following the rule that the heftier the adult bear, the more domineering. Besides their sheer size, the males enforced their place in the hierarchy through body posturing and head movements. Assertive bears tend to stare, roar, or "mouth" – wave their heads with mouth open near the head of another bear. Usually, these threats are bluffs. Nonetheless, the subordinate bear indicates its submission by easing off and giving the dominant animal more space. By maintaining some distance from one another and being observant of space rights, serious conflicts were avoided.

Females with cubs were next in the hierarchy. Not surprisingly, sows weighed the costs and benefits of proximity to one another, or to males, with caution. At the bottom of the hierarchy came immature bears, which were forced to fit in. They avoided encounters by being attentive.

Reflecting on our Katmai experience, as exciting as it was to see grizzlies so tolerant of humans, we were left with a nagging concern. Were the bears getting the wrong impression of us – benign, binocular-bearing, respectful of their space? Disappointingly, grizzlies are still shot for "sport" – which is officially sanctioned and even encouraged by a legal hunting season – or are treated as "nuisance" animals. The entire Katmai peninsula is not national park. Some of the adjacent land is "national preserve," where grizzly hunting happens, and is even promoted. Would some of these trusting bears misjudge their next human encounter?

The home range system employed by grizzly bears is less common in animals than territorial behaviour involving more active aggression and defence. Many animals expend the energy and take the required risks to defend a territory, because it is usually a more successful way of dealing with land ownership. Territory-holders get to know their neighbours, with whom they quickly work out disputes, and so they seldom need to compete directly with them. In comparison, animals in a home range situation may be forced into repeated confrontations with many members of the same species – anyone who wanders by – if such competitors are not more vigorously repelled.[7]

Why, then, would bears and certain other species have home ranges rather than territories? The answer lies in the observation that most home ranges and home range species are large. Large-bodied species need large amounts of land to obtain sufficient resources, which places heavy travel demands on them. Travelling around to all corners of one's domain to expel intruders may just demand too much energy to be a viable strategy.

The biological script and procedural policies for territorial behaviour is complicated, because territories come in many forms. Different species adjust their territorial systems in response to various ecological situations. Some are held all year, others only seasonally. Hawks, with widely dispersed prey, commonly defend only a nest site. So do colonial birds, which nest in large aggregations as a defence against predators but feed elsewhere with other members of the colony, competitively, if food is scarce. Group territories are claimed by social units like packs of wolves or prides of lions, and "within-pecking-distance" territories by colonial nesting seabirds such as gulls and penguins. Some species patrol and mark their boundaries; others, like ruffed grouse, defend only an "invincible centre."

Central to territorial behaviour is aggression, a topic that has attracted considerable study, maybe because humans can identify so well with it. Aggression is inherent in the neural circuits of the brain. But normally it is aggression with a caveat. It is safer for an animal to defend its territory through posture, bluff, and ferocious appearance than to attack and risk being killed. Animal songs and calls, too, are often expressions of ritualized aggression.

Some expressions of aggression are even more subtle, such as the direct stare of the brown bears in the Katmai, or the tail or body positions of social animals such as wolves. Often, the territory-holder tries to intimidate the trespasser with its display; mammals bare fangs, raise ears and hair, while birds lift their crests, fluff body feathers, spread and wave their wings and tails. Only if intimidation displays fail is the territory owner forced to attack and chase the intruder.

Consensus holds that protecting a food resource, securing and keeping a mate, and defending the young are the primary reasons for territorial behaviour, and much has been written about their relative importance. But the bulk of evidence for territoriality is related to food, particularly driven by competition with other members of the same species.[8] Feast and famine determine the size of a territory and the destiny of its stakeholder, a generalization that is true not only for the large and charismatic but also for the small and pedestrian.[9]

The best evidence for the imperative of food is that territory size, like a supple, expandable disk, increases or decreases with its availability, thereby influencing the density of the population. A classic example of this "elastic territory" was shown by dunlins, small shorebirds that nest in Alaska. In areas where food was abundant, dunlin numbers were high and their territories were small. Under pressure from high density, these territories diminished until they assumed a polygonal or hexagonal shape, like the shape of cells in a honeycomb, minimizing the perimeter of the boundary. In contrast, in a region where food was less abundant, dunlins were less dense and individual territories were larger and more variable in shape.[10]

This malleable feature of territories in response to food has been observed for many other species, including female red squirrels in northern English forests.[11] Similarly, eastern chipmunks in Pennsylvania decreased their territory size when food levels increased, and their population enlarged as well because of an inflow from neighbouring areas.[12] Studies of ovenbirds in Ontario also showed a decline in territory size when food increased.[13] Hummingbirds and Hawaiian honeycreepers normally defend nectar regions vigorously, but when resources are prolific and food is plentiful for all, their territories are not defended.[14]

What if the importance of real estate were not so much to secure food, but to attract and hold a mate? Evidence exists for that, too, demonstrated particularly by the use of song. Although song is commonly considered part of passive territorial defence, Donald Kroodsma, who has spent his career studying birdsong, disagrees. "I would argue that the male-male singing contests in spring are less about territorial rights than about impressing females; the abundant spring song is more about sex than territory, and if there were no sex, there would be no song."[15] He asks about mockingbirds, one of nature's great singers, "And how can the male best impress? Is it by how much of the day he sings . . . or is it a sprint to see how much he can sing within a short time . . . or does she value the quality of his mimicry, or the quantity of it, or his overall repertoire size? Exactly how he impresses is known only by the female mockingbird."[16]

Kroodsma's unexpected contention provided us with an excuse for a trip to the tropics. For migrant birds from the north overwintering in the tropics, there is no sex. But would there be territorial behaviour and song? Where so many species live – both migrants and permanent residents – we expected both to be intense. About half of all land birds that breed in North America overwinter in the neotropics (tropical areas of Mexico, the Caribbean, Central America, Cuba, and northern South America), and in doing so they approximately double the abundance of birds that live there. As well, the migratory birds alone are packed in more closely than on their North American breeding grounds; slightly more than 2 million square kilometres of wintering grounds must accommodate birds that migrate from breeding grounds more than eight times as large.[17] Compounding the space problem, too, is the fact that the amount of usable land in the tropics is shrinking as humans expand cities, towns, and agriculture.

Early one March we travelled to Belize. Our destination was Las Cuevos Biological Research Station in the south-central part of the country. The station, a cluster of long, wooden buildings perched on four-metre (twelve-foot) stilts, was situated in a grassy clearing wreathed by tropical forest. The buildings provided lodging not only for a handful of scientists but also for the local inhabitants who ran and maintained the facility, and a contingent of armed guards who protected the border from Guatemalan encroachers. This was the only time we have bird-watched accompanied by lookouts carrying AK-47s.

On our first morning at Las Cuevos, we rose before daylight to record what we expected would be a resounding dawn chorus. Standing at the edge of the forest with our microphone and minidisc in hand, we waited for the concert to begin. The first sounds to break the stillness were the mournful, hollow, repetitive "whoo whoo"s of white-tipped doves. Synchronized with the dove floated the velvet, owl-like tones of a blue-crowned motmot, followed by emotive, high-pitched whistles of several greater tinamous. The haunting calls of these three opening singers seemed more mystical than real as they wafted through the sound pathways of the jungle. Soon, other birds joined in: cascading, bell-like notes of blue-black grosbeaks, hoarse calls of scarlet macaws, and the squawky, discordant notes of red-lored parrots. Between these bird calls, however, was a lot of silence. We waited for a full crescendo to open up, but it never did. A few more species – barred antshrike, melodious blackbird, Montezuma oropendola – chimed in, but compared with the dawn chorus in the eastern Canadian forests, the tropical forest was hushed.

Later that morning we walked the forest trails with an experienced bird guide, who was adept at seeing motionless birds. It soon became apparent that a rich diversity of species was actually there. In fact, more bird species inhabit the neotropics than any other region in the world. Some entire families of birds are confined to this region,[18] and a rich literature attests to the commonness of their territorial behaviour. Why the paradox – lots of territorial species but little song?

Part of the answer relates to the extended length of the breeding season. In an open thornbush forest at Crooked Tree Sanctuary in northern Belize, a curious rufous-tailed hummingbird zipped past our heads. Its brilliant, prismatic colour directed our attention to a dense thicket, where a frenzied blur of movement revealed its mate assiduously preparing its nest. But only a few days earlier, we had followed the flash of a vermilion flycatcher to its more featureless mate, discreetly ensconced in her nest, already incubating eggs. And towards evening that same day, a wood stork flew from the edge of a thickly wooded bayou, where it had been feeding, to the top of a palm. Sitting stolidly in a twig stronghold, two fledglings stretched their necks and clamoured to be fed. The significance here was one of wide latitude in the timing of breeding, a characteristic that many researchers have noted. Not only do various species breed at different times, but the breeding seasons for individual species may be as long as four to eight months, timed to coincide with the prolonged abundance of food – fruit, nectar, insects, seeds.[19]

Besides a protracted breeding season, several other characteristics of territoriality in the tropics reduce the amount of song. The year-long defence of a territory is common, to accommodate all-season food requirements of tropical birds, resulting in larger territories – roughly ten times the size of those in temperate forests. Large territories mean that the nearest neighbour of the same species is often far away, too far to engage in bouts of singing.[20]

Reducing birdsong in the tropics, as well, is a tendency for limited competition within species. Resident tropical birds usually experience low over-winter mortality, partly the result of not having to undergo the perils of migration, which results in fewer territorial vacancies, and in turn reduces the need for musical want-ads to attract a mate. As well, with many tropical species, such as dusky antbirds, floaters searching for territories are rare since young birds tend to stay put until the next breeding season, offering little competition for land.[21]

A final reason for this low amount of song is that during the protracted breeding season, nesting birds become ever more furtive and keep quiet since predation on eggs and young is so high.[22] Predators are common among the high diversity of tropical birds.

In contrast, in North American temperate forests the musical score is much more dramatic. Competition is intense for land, and equally intense for mates. Birds must breed quickly, a situation forced on them by the short season, with 90 percent of breeding and rearing taking place within a few weeks. In addition, high over-winter mortality, partly caused by migration, results in unoccupied areas available when migrants return in spring. New birds may constitute as much as 50 percent of a bird's neighbours.[23]

What about the temperate forest migrants when they are wintering in the tropics? How do they fit in? One morning at Los Cuevos, while scanning through a mixed flock of tropical species that were livening up a huge ceiba tree, we suddenly spotted a familiar bird, a black-throated green warbler. Then a Canada warbler, and more: Wilson's warbler, yellow warbler, indigo bunting. It was like watching the hometown hockey team in an away game – without any soundtrack. They were all going about their business in silence.

Biologists who study these wintering birds have come up with some broadly accepted generalizations. On the wintering grounds, neotropical migrants are accommodated with apparent ease. They just slip in, normally exploiting significantly different niches than the resident tropical birds.[24] The neotropics is the migrants' home for more than half the year, since this is where they evolved, alongside the species that stay there year round. So, for the migrants, their niches are waiting for them when they return from their breeding grounds.[25] Most take advantage of that fact either by re-establishing their territory or by carving out a new one in appropriate habitat.[26]

For these migrants from temperate forests, then, territoriality on their wintering grounds is clearly not related to breeding. Moreover, not only are they indifferent to a future mate, but the males of many species actively repel the females, causing the females to set up their own, independent territories. In species where the males and females have distinctly different plumages, such as the American Redstart, males appear to grab the best territories and force females to find theirs, often in submarginal habitat.[27] However, this "mean streak" is less easily accomplished when the plumages of both sexes are identical, as in most fly-catchers. Then, the females manage to claim territories of equally high quality.

And song – it is rarely used except just before migration, when it is triggered by hormonal changes related to the upcoming breeding season. Instead, winter territories are maintained by simple notes or other sounds,[28] or in silence.

This observation that birds do not need song to successfully defend a territory appears to support the notion that, during the breeding season, temperate forest birds use song more for sex – to attract a mate – than for defending a territory, as Kroodsma contended. But why must it be one or the other? More likely, song serves a dual function during the breeding season, with the ability to defend a territory being a component of mate attraction and breeding success. How else can we interpret the vigorous defence of territories, accompanied by vigorous singing, that birds display when we play the songs of other males that the birds assume are trespassers (see Chapter 3)? In behavioural ecology, however, attributing motive is a dangerous pastime.

Home range, hierarchy, territory – one more form of spacing behaviour is commonly found, not only in the tropics but also in temperate regions, though to a much lesser extent. It, too, significantly affects birdsong. It is the absence of territoriality, and it happens in response to a clumped and uncertain food supply. Then, the logical alternative is to become nomadic.[29] In temperate regions, birds whose specialized diet consists mostly or entirely of tree seeds are typical candidates for nomadism: crossbills, grosbeaks, and Clark's nutcracker. Conifer trees have good crop years and bad, here or there, and these bird species course the country looking for productive stands.

In the tropics, the primary nomadic species are fruit-eaters. When a flock of screaming parrots or scarlet macaws flies by, the forest resonates with their calls. The fruit they seek is so abundant locally that there is little competition for it.[30] But the fruiting trees are widely spread and the parrots must search broadly, their calls helping to keep them together. Then, like a thunderbolt, they are gone, leaving the forest in shocked stillness.

Insect-feeding birds in the tropics, as well as many birds during migration, also establish nomadic flocks often comprised of several species. They, too, sweep through the forest and depart. Most species in a mixed flock have their own unique prey and techniques of capture, which reduces competition. For instance, various species of bark- and foliage-gleaners partition the forest vertically, scrutinizing the vegetation for insects at different heights.[31]

Pursuing a nomadic lifestyle in flocks has the reward of increased hunting efficiency for an individual, by reducing search time in areas already explored by others and by having more eyes to find food. But it comes with a personal dilemma. In a flock, is it better to call when you find food, or be silent?

A determining factor seems to be the relative priority of finding versus sharing food once it has been found – again, a consideration of optimum foraging strategy. If your flock rarely finds food, but you do, then remaining silent is most beneficial to you; however, if food is abundant, then offering to share is inconsequential. In both cases, however, flocking would be of no particular advantage and would disappear.

However, an overriding benefit exists both for flocking and advertising your successes. Calling, while appearing detrimental to the individual who must then share its discoveries, falls into a class of behaviour named mutual altruism – you scratch my back and I will scratch yours. The next food patch may be discovered not by you but by another flock member, and you will benefit reciprocally, just as it did when you found the food source and called. Much seemingly co-operative behaviour in animals is really mutual altruism – in humans, too.

Flock nomadism has another cost-benefit consideration: the cost is being conspicuous to predators; the benefit is early predator detection. One study found that wood pigeons spotted goshawks more quickly when they were members of a flock, and could escape by getting up their flying speed.[32] As well, in both flock and herd situations, the principle of safety in numbers aids the individual, because a predator can be deterred or confused when attacking a group. And the odds of being a victim decrease in a crowd, since the simple probability of being killed goes down with the number of other victims available. As well, a group can offer defences that single animals cannot. We have watched elk bunch up to screen juveniles or weak animals, wildebeest hide their young from marauding hyenas by passing them to the safe side of the herd, and muskoxen huddle together in a defensive ring when a pack of arctic wolves approached. Since many predators select the young or the weak, these individuals are less susceptible within the confines of a group.[33]

Altogether, flocking and herding, worked over carefully by natural selection to include mutual altruism, group defence, predator swamping, and mixed flock niche-separation, are effective spacing strategies for many species.

Lek behaviour, in which several males display on a patch of ground to attract a mate, is a less common form of spacing behaviour, but stands out as flashy and flamboyant. It is found in the tropics among tiny, brightly coloured manakin birds, in the Arctic in some species of shorebirds, on the African savannahs with species of antelopes, and on North American grasslands with grouse. These groups have little in common, nor do they have close relatives that exhibit lek behaviour, making the reasons for it obscure.

The sandhills of western Nebraska are one of the last strongholds for the greater prairie chicken, a lek species that once populated the plains. Prairie chickens have declined in North America and at present are living on borrowed time. We travelled to the small farming town of Mullen, Nebraska, one of their remaining safe havens, to see them.

At 4:00 a.m., Mitch, the owner of a small, 1950s-vintage motel, picked us up in his battered Suburban and drove us to a grassland amphitheatre to watch the prairie chickens' timeless performance. On the way, he explained how he had taken an old, surplus school bus to act as a blind and moved it closer to the display grounds each afternoon when the birds were not around. Finally he had the bus right on the edge of the grounds, a slightly elevated knoll covered with short, sparse grasses. The performers completely accepted this unnatural object.

It was still dark when we arrived and took front seats for the "chicken dance." We set up our scopes, cameras, and sound equipment by the windows, and waited. At first diffused light, the whirring of wings announced the arrival of the males, about thirty overall, which flew singly or in small groups to the mound. Then came the first "booming" – three notes in quick succession, sounding like air blown across the top of a bottle. As the light improved, the performers materialized, their heads lowered, wings spread, and tails raised. Tufts of neck-feathers that looked like upright ears rose to expose two inflated, orange esophageal air sacs. Then, stamping their feet rapidly, the birds rushed across the dancing grounds, repeatedly issuing their three booming tones. Each male had the same characteristic suite of moves in this primeval dance – beating the air with his wings, vibrating his tail, booming, whirling, dancing, and charging the others with aerial leaps.

Soon two females appeared and walked through the core of the lek. The displays became more vigorous, and the chorus of rhythmic hoots intensified. The females, however, seemed unimpressed – still shopping around. By 9:00 a.m., the energy of the dancers waned and the birds quieted down. One by one, or in small groups, the males flew off in different directions, and the show was over.

Sex motivates prairie chickens to assemble in the spring at these communal arenas, where males advertise for a winning ticket in the reproduction raffle. Each female has her own piece of land somewhere nearby, more of a home range than a defended territory, chosen because of nesting cover. But the males have no interest in land beyond the lek. Being primarily nomadic, they stay close enough to the showground to fly in each morning.

Breeding success for the males depends on holding the choicest real estate – the centre of the lek. At the beginning of the breeding season, each male confines his movements to the best portion in the arena he can secure and returns there each morning to defend his mini-territory. The male's objective is to claim a central position, because that is the area of interest to the females. Consequently, only 10 percent of the males do 75 percent or more of the mating. As long as a few high-quality males are present to breed with the females, the rest are superfluous.

The females do the choosing. The lek allows them to ascertain male fitness.[34] The females dictate who will sire their offspring and thus the genetic fate of the males.[35] But then, who gets selected or dumped? Does the female select the male on the basis of his show, the "hotshot" model? Or does she select the male on the basis of the real estate he holds, the "hotspot" model?[36] Convincing evidence to separate these two possibilities is difficult to obtain. Perhaps, it is both.

Other than in zoos, all species operate as populations – interacting individuals of the same species with common needs. Spacing systems represent the rules of competitive engagement that match a species' biology to its ecosystem. They have been carefully scrutinized by natural selection, because they are the most basic of population requirements. They are time-tested to make adaptive sense.

Spacing systems need space – less for small species, plenty for the big ones. They are the property-multiplier of population numbers. But land for wild things is diminishing. Populations of grizzlies, prairie chickens, many neotropical migrants, and myriad other species, all have felt the crunch of diminishing land.

Conservation is not a small-scale land game. It means protecting populations – whole and intact, large enough for genetic variability and ecological functions. And it means giving species enough room for their space needs. There is a pattern in the intricate dovetailing of species and the room to exist that they need – a balance, an ingenious meshing of the two. We cannot relegate nature only to leftover land.

CHAPTER SIX

HIGH-STAKES LIVING

At times, population rules appear to be on hold. Prodigious numbers of individuals mass together in one place. Spacing systems seem to have failed, order been abandoned. Or are these wildlife aggregations the result of other rules? Do they reflect inherent biological strength? Does the principle of strength in numbers assure a more certain future?

WE FOUND THEM IN LAVISH NUMBERS on sandbars, in grain fields, on the wind etched against a burnished sky – contingents of sandhill cranes, legs and necks outstretched, their trumpeting calls saturating the air. They were at the Audubon Society's Rowe Sanctuary on the Platte River near Kearney, Nebraska, situated on the central flyway of North America near the geographical centre of the continent. As though passing through the neck of an hourglass, the birds funnel there from a wide wintering distribution farther south, then fan out again as they continue north.

For centuries, thousands of sandhill cranes have assembled on the gravel beds of the Platte River, fuelling up on their northward spring migration.[1] This site is the most important stopover for cranes in North America, their numbers building to 400,000 or 500,000 in late March each year.

Why there? The Platte is broad, braided, and shallow – ideal for cranes. Here they wait for their breeding grounds farther north to thaw. Here are wetlands, hayfields, pastures, and the stubble of surrounding cornfields where they can fatten up for the final leg of their migration. Here they can rest, court, and roost on midstream sandbars, safe from predators.

We were met by Bud, a gaunt, friendly Nebraskan in a coonskin cap, an employee of the sanctuary. Animatedly, he pointed out sandhills feeding in the distant cornfields. Other birds, grey as winter clouds, drifted down and landed on the remnants of last year's crop, probing the ground for grain. Cranes circled overhead; flock after flock flew back and forth, clamouring constantly.

We threw our gear into Bud's dust-covered truck. He wanted to get us to our riverside blind before the 5:00 p.m. deadline, ahead of the incoming roosting cranes. The knee-high blind, an oversized wooden casket on the water's edge, was big enough for us, our sleeping bags, tripods, a "honey pot," and not much more. We would stay there until Bud retrieved us the next morning, after the cranes had left the river flats to forage in the fields.

A century and a half ago, the prairies of Nebraska stretched on and on in a sea of grass. Back then, there was permanence to the landscape. The arrival of Europeans changed all that, as the Platte River Valley, part of the famed Oregon Trail, became a major westward highway for the pioneers. Today, the Platte cuts a sinuous path across a flat agricultural landscape that produces mainly corn. Now, huge irrigation machines scribe circle patterns on the land, visible even from high-flying aircraft. Railroads, Interstate 80, and smaller right-of-ways crisscross the state.

"Too thick to drink and too thin to plough," the pioneers declared disparagingly. But the Platte's ankle-deep profile was just right for sandhill cranes. They are fastidious in their choice of roosting habitat, avoiding confined locations that may make them vulnerable to predation. Over the centuries, the mile-wide river has braided and inter-braided erratically, changing and remoulding gravel bars with the whims of time, never allowing its sandy shoals to run deep. Each spring, during breakup, the river scoured the vegetation on the sandbars, uprooting small willows along its banks. Spring floods submerged adjacent sedge meadows, helping to decompose dead material and provide a banquet of snails and other high-protein invertebrates for the cranes.

That evening, as on all spring evenings, bugle-like calls heralded the sandhills' arrival. The setting sun had signalled the birds to leave the cornfields and return to their roosts on the protective river gravels. Long, snaking lines of cranes streaked the sky. Gracefully they turned in unison, calling out in their distinctive quavering voices. Through binoculars, we watched their silhouettes drift down against the sunset onto the sandbars.

Seconds later, a car spun its wheels on a distant road and raced away, shattering the silence. Pandemonium broke loose. The river bars detonated with thousands of sandhills surging back into the sky, calling out in alarm. Column after column, the winged swarm lifted off for safety. Soon the stretch of river was empty. Nothing could have prepared us for this. We knelt there, dazed and dejected.

But before long, in the fading light, the cranes swept back. Where else could they safely go? Their deep, rolling "k-r-r-roo" carried through the still air as the flocks sailed over us and alighted again. The flow of birds was constant. More and more dark forms descended like parachutes; the banks of the Platte filled with sand-hills and resounded with crane music. One bird to our right began to dance, and soon the contagion spread. These dancing duets can begin at any time. Spring mating ushers in this ritualized behaviour: mated pairs, with heads bobbing and wings drooping, bow, prance, and leap into the air.

Eventually, darkness shrouded the forest of slender necks. But the night was not silent. Dozing in our sleeping bags, we listened to their various calls, rendered more superb and mellow by the velvet night.

As morning approached, we flopped on all fours to the openings in the blind. Through wisps of mist, a faint light penetrated the eastern horizon, outlining obscure shapes standing motionless on the shoals. As the light improved we could make out cranes along the shallows in all directions. Most had their heads tucked back with their bills slipped into their scapulars.

Gradually, they became more active; some ruffled their plumage with vigorous shrugs, preened, and strode stiffly around. Then the flock began to drift downriver. Periodically a bird leaped up into the air, beginning an explosion of bows and prances. No wonder the early settlers called them "preacher birds" – perhaps preachers were more entertaining back then.

Today, however, the Platte is no longer the river of the past. Arrested by reservoirs, it has become seriously degraded. Dams in Colorado, Wyoming, and Nebraska suppress its historic flow by 70 percent. Without the river's annual flooding, vegetation has encroached onto the gravel beds and banks. The roosting sites are less suitable as predator-free zones. Less and less do they fit the genetically programmed criteria of suitable habitat.

How much environmental change will the cranes tolerate? Wary of humans and disturbance, might the birds continue on their migration before they have fattened up and rested? Someday, might they simply just abandon the Platte?

But so far, the cranes keep coming. What, then, is the lure? The answer – a familiar one – is food! At present, its abundance out-weighs the impact of changes to the river. Cornfields and grain are replacing the vast sedge meadows and the cranes' invertebrate diet. Under this new-world order, the cranes have adapted to and profited from this human-caused amendment and, at present, are the big winners. Their populations have expanded. The flocks feed and feed again, accumulating the life energies that will power them northwards, building up their fat reserves, preparing for the great energy drain ahead. During migration, adult cranes need to accu-mulate fat reserves of about 10 percent of their body weight – almost half a kilogram or one pound (that is roughly the equiva-lent of a 70-kilogram or 150-pound human gaining 7 kilograms or 15 pounds in a month). For half of the population that travels to tundra nesting grounds there will be little to eat for several weeks until spring arrives in the north.

Before much of North America was brought under this culti-vation makeover, crane numbers would have been limited by the availability of spring food. That limitation no longer holds. Today, no density-dependent feedback dampens the cranes' population growth. Their numbers continue to increase, tempered neither by increasing mortality nor by decreasing fecundity. At least not yet. But, in the long run, will cranes be the losers? A century ago, a 320-kilometre stretch of the Platte provided roosting habitat for cranes. In recent times, they have become concentrated into only 128 kilometres.

No species can expand in numbers indefinitely. Eventually, negative feedback will halt its population growth. Rapid growth, such as the cranes have experienced, often results in overshooting the carrying capacity of their environment and leads to rapid decline. Could any further crowding on the Platte be disastrous, predisposing the cranes to disease? Already, diseases such as avian cholera have taken a toll on cranes and other waterfowl in areas that are crowded and stressed.

Eventually, their vital staging grounds on a drying-up river might fail to support the cranes' vast numbers, and their passion for corn might be their downfall. Have they put all their eggs in one basket? As we stood listening to their resonating cries, we wondered – how long?

Near Vancouver one spring, we watched a rippling upsurge of white against a blue sky. Snow geese were streaming overhead, luminescent in the sun's rays. The air overflowed with their soft, melodious gabbling. Some mounted into the sky, others poured onto fields of winter wheat, a flurry of geese rising and falling.

From a road edge, we watched the display. They grazed intently for a while, then, with an eruption of white, surged skyward again as one expansive wave, twisting and turning in the wind. We tried to estimate their numbers, starting at one end of the wave and then the other. Several thousand made up the throng.

Wild migratory geese hold a peculiar fascination for many people; it is difficult to know just why. Is it the sudden burst of wild honking that evokes a memory of remote, untamed places? Could it be their sheer abundance? Or is it their coordinated movements, their unchoreographed precision, their neat V chevrons slicing the sky?

As we drove to the George Reifel Migratory Bird Sanctuary that March day, marshes and fields were ringing with the clamour of widgeons, mallards, swans – trumpeter and tundra – pintails, gadwalls, and green-winged teal. But best of all were the wintering snow geese massed in the tens of thousands in various places across the Fraser Delta. Here on their winter range, from October until March, snow geese press into wildlife refuges and farm fields. But they are only visitors. Soon the great white flotillas would abandon these productive, wintering wetlands and migrate along their ancestral route to nest on the plateaus of remote Wrangel Island in the Chukchi Sea north of Siberia.

The Fraser River estuary is a great biological factory that creates a nutrient-rich environment for an immense number of species. Its tidal marshes and the adjacent farmlands in southwest British Columbia and northwest Washington State provide one of the most important wintering waterfowl habitats on the West Coast.[2] In the distance is the sprawl of Vancouver's suburbs and industrial developments. High towers spike the sky. Beyond the dikes and marshes, barges, oil tankers, ferries, and pleasure boats jostle back and forth. However, in the tidal flats, on croplands, and in sewage lagoons, the snow geese are completely tolerant of their diverse human-altered environment, which once was washed by the sea but now is protected by dikes.

The story for most North American snow geese is one of both misfortune and resurgence. In 1535, Jacques Cartier reported "many thousands of white and grey geese" in the St. Lawrence Valley, apparently seeing both the blue and white colour morphs.[3] With early settlement, however, came over-hunting, and populations declined. By the early 1900s, snow geese were rare. Then, times changed. North American snow goose populations have exploded, especially since 1960. Now, breeding snow geese are one of the most abundant species of waterfowl in the world. The mid-continental population has risen to around 6 million and is increasing by approximately 5 percent a year, causing considerable concern for wildlife managers. By 1992, the Wrangel Island population that winters in the Vancouver's Fraser Delta had increased to 70,000 birds. Today, along the St. Lawrence River, snow geese whiten the fields in spring. At Bombay Hook Refuge, in Delaware, 150,000 take up residence for the winter, digging up cordgrass on the tidal flats or nibbling at the crop remains in the fields.[4]

Like sandhill cranes, snow geese are taking advantage of previously unavailable food and have made the shift effectively from marshes to fields for much of their feeding. The widespread availability of grain crops and recent declines in hunting have reduced winter mortality, allowing the birds to become over-efficient breeding machines. Before European settlement, snow geese fed on fibrous marsh vegetation, grubbing rhizomes and grazing young shoots. Starvation thinned them out any time their population numbers increased beyond the carrying capacity of the marshes. Weakened geese raised fewer chicks. But increasingly, humans drain marshland for soybeans, corn, rice, and a whole host of other crops, which for snow geese hold a "fast food" appeal.[5]

And so, the carrying capacity of southern marshlands that regulated snow goose populations prior to settlement times no longer applies. Consequently, the geese are overwhelming their tundra nesting grounds. Like humans, they are marring, overpopulating, and consuming their environment. Massive numbers of the mid-continental snow goose population breed on the great muskeg of the Hudson Bay lowlands. These extensive wetlands and peatlands are vital to the thriving geese, offering them a profusion of sedges and grasses where they can feed and proliferate. But their sheer numbers have made excessive grubbing behaviour mal-adaptive, and they are destroying their fragile lowland habitat. Their intense rifling damages the plant cover on the inorganic soil, creating a muddy mess. The soil desiccates, becomes hypersaline, and ultimately turns into a crusted hardpan. The geese move on to other productive areas, and the process begins again. By 1993, snow geese had damaged 65 percent of the intertidal vegetation of La Perouse Bay, near Cape Churchill, Manitoba. At Karrak Lake, Alaska, they had destroyed 59 percent of the plant cover.[6] Habitat recovery in these places is doubtful, especially with con-tinued soaring goose numbers.

And now there is a new threat, a recent discovery that fresh-water lakes and ponds in the southern part of the permafrost zone are draining away into subsurface soils. Satellite imagery has doc-umented this wetland loss due to climate warming over the last three decades. These ponds and their islands provide important predator escape, in addition to food, for the nesting geese and their young broods.[7]

Can their breeding range endure these mounting impacts? Someday, for snow geese as for sandhills, will their habitat have the last say?

But on that March day, we wanted to be optimistic about what the future held for snow geese. They were a dazzling sight: albescent except for black wing tips; long, tapered feathers cas-cading down their backs. On that day, they brought a wildness to the Reifel dikes.

Another year, and on the other side of the world, we were in Africa to study wild Cape hunting dogs with Kathy Alexander, a biologist in Kenya's Masai Mara National Reserve. The Mara is a vast haven for wildlife in the remote, sparsely inhabited southwest part of Kenya, nestled against the Tanzanian border. The population of Cape hunting dogs, however, had been decimated by disease, which they had apparently contracted from the Masai's domestic dogs. So, unable to study the wild dogs, we monitored some radio-collared jackals for Kathy. Each day, we left the tent-lodge at first light in her Land Rover, driving to the ridge tops to pick up their radio signals. Sometimes, we took members of the lodge's African staff with us. Some had never seen cheetahs, lions, or leopards – their own birthright. And with more free time than we had expected, we also tried to understand the relationship between herbivores and predators on the Mara, which has adapted to an extensive human presence over a long time.

East Africa is a realm apart for its multitude of magnificent animals. The grasslands form a landscape with endless vistas that provide the stage for a variety of dramatic events. Mara's wilderness is inextricably linked with the ebb and flow of the seasons, the luck of the predators, a superabundance of prey, and a spectacular migration of ungulate herds, principally vast numbers of wildebeest.

When we arrived at the Mara, the savannah landscape was alive with wildebeest. Each day we watched them from our Land Rover, partially concealed in a grove of acacia trees. Their numbers stretched endlessly across the plains and culminated in a huge throng crossing the Mara River.

Annual rains trigger more than a million wildebeest to forsake the dry plains of the Serengeti and journey northward, lured by the scent of moisture and the Mara's abundant new grass. But the Mara River forms the wildebeest's biggest obstacle. Each day, we watched their numbers building on the far side as animals reached the water then turned back, reluctant to take the plunge. More and more kept coming, and finally, pushed by the mob, the desperate animals in front plunged in. They crossed the river, thrashing wretchedly in the current, then scaled the precipitous clay bank, some falling back into the water. In surges, more wildebeest leaped in, swam for their lives, then struggled heroically to reach the plains above. Some unfortunate ones were speared by flailing legs; others were injured on rocks and floated downstream. Downriver, wildebeest carcasses lined the banks while gratified undertakers – crocodiles, jackals, and vultures – cleaned up.

Most wildebeest, however, made it safely across. On our side they grazed and gambolled in a black swirl across the plains. Their unremitting grunts filled the early mornings and evenings. Birds of prey – marabous, martial eagles, and kestrels – patrolled the skies. For them, this was the season of plenty. So it was, too, for lion, cheetah, leopard, hyena, jackal. Little would go to waste. The wildebeest females would drop their calves all within a few days, synchronized by natural selection to overwhelm prospective predators. Still, plenty would die. Of half a million wildebeest calves born, tens of thousands would succumb to predation, starvation, or disease – life on the African savannah is problematical.[8]

Nonetheless, it is the availability of forage that keeps the wildebeest population at more than a million. That forage is a function of seasonal rains – days of soft summer drizzle that whispers in the grasses, or sudden deluges that whip the scattered acacia trees into a frenzy. Stimulating the grass to grow, too, is the grazing, trampling, and manuring not only of wildebeest but of Thompson's and Grant's gazelles, zebra, impalas, and giraffes. Together, but each species in its own time-defined niche, they gradually strip the plains, zebras grazing on the rough tallest part of the grass, wildebeest cropping the shorter grass that the zebras expose, then Thompson's gazelles eating the regrowth.[9] Grasses have been stimulated by grazing all over the world through evolutionary time, whether grazed by snow geese, cranes, or wildebeest.

Large tracts of land are essential for this multitude. Free movement of animals in vast landscapes assures genetic flow, thereby maintaining viability and health. It gives animals options to move from areas of drought to regions where rains have re-greened the plains.

Historically, the wildebeest from Tanzania flowed beyond the Mara and mingled with the Loita Plains population to the east. In recent years, much of the Loita Plains have been ploughed up to grow wheat. Now, fields of ripening grain stretch to the horizon, obliterating the native landscape once used by the herbivores. Wildlife pays no attention to unfenced boundaries, and many animals are shot when they wander out of parks or reserves and trespass onto agricultural lands. Wildlife habitat has been lost, as well, to strawberries and flowers for European markets, or overgrazed by exotic mammals such as goats and cattle. The result is wildlife in a squeeze. This forced concentration may alter the delicate predator-prey balances that have persisted for so long. Being constrained and concentrated in this way also encourages outbreaks of epidemic diseases and hinders populations with reduced genetic variablility.[10] Animals without habitat are simply singularities biding their time until extinction.

And so, the immense wildebeest numbers decline steadily in these ever-shrinking and ever-changing habitats. Like the white and black rhinos that once numbered in the thousands but are in danger of extinction today, will we see wildebeest numbers simply collapse? In another fifty years will their great abundance still be on the move, or will we find only remnant herds standing around in small wildlife preserves?

Over a week-long period, we trekked through thirteen reserves in central Mexico's Sierra Madre mountains with John's graduate student, Jurgen Hoeth, asking what was important to the survival of monarch butterflies in their overwintering roost sites found there. Jurgen knew monarch butterflies and their reserves well, having grown up in Mexico, and wanted to explore the extent to which tree cutting might be acceptable in the monarch habitat. In the local villages, wood was essential for cooking and warmth, and strict forest protection for the butterflies had prompted one community to clear-cut a monarch reserve in both protest and desperation. So, there was a controversy and a conservation issue to be solved. Was careful, limited cutting a possible answer?

No butterfly is better known than the monarch, famous for its "miracle migration" across North America. Equipped with their own internal, luxury GPS, monarchs fly an astonishing 6,000-kilometre journey south from their most northern natal locales along established routes to their wintering sites, to which they have a strong, genetically endowed fidelity. The southward journey is through totally unfamiliar territory, as it is undertaken not by survivors of the previous winter but by their descendants, up to five generations later. Yet these descendants find their way back to the same reserve, and even the same natal trees used by their great-great-great-grandparents.

At dawn we entered the monarch reserve. Beams of sunlight penetrating from breaches in the canopy outlined the ghostlike oyamel firs swathed with butterflies – a perfect ecological integration between species and their environment. The limbs of the firs, heavy with the dense assemblages of monarchs, quivered slowly as the sun's rays touched each cluster. Butterfly after butterfly opened and closed its wings, gradually responding to the energizing warmth. Before long, a blizzard of butterflies spilled out above us, falling like immense snowflakes – a cascade of opulence – saturating the air. We could scarcely move for fear of stepping on those that, not fully warm, had floated to the ground. All around us, they sailed slowly past, indifferent to us, flying towards some light-green shrubs that proffered a profusion of yellow and white flowers, and sustenance, and life.

For centuries, disturbance must have been minimal in the old stands of oyamel forests where monarchs wintered. Temperatures at the chosen altitudes were severe enough to numb the butterflies into dormancy but not to kill them. But now, when humans cut the forests at those crucial elevations, monarchs are forced to aggregate elsewhere. If they go higher or lower, their fate is sealed.

We located one such unfortunate colony, after discovering that its traditional site had been logged. The butterflies had moved up-mountain, and a winter storm had frozen them in their unsuitable new locale. In their crypt, we found their wings in the millions, covering the ground many layers deep. Anywhere their frozen corpses had fallen, mice had consumed the butterflies' fat-rich abdomens.

Monarchs are vulnerable on migration, too. They are a highly specialized, genetically programmed species, intimately integrated into the communities they inhabit. They orchestrate their northward departure in spring with the emergence of milkweed, the only food suitable for their larvae. In what ways will global warming affect that timing? Might climate change disrupt relationships that now exist between the use of stopover areas and the availability of food? Maybe masses of monarchs with intercontinental demands are historical anomalies in today's world.

In all these places – the Platte, the Fraser delta, the Mara plains, the Sierra Madre mountains – key environmental and biological variables have lined up to make massed superabundance possible: a high rate of reproduction and survival of young, an absorbable rate of mortality, a concentration of food, a supportive habitat, an abandonment of territoriality or other extensive spacing systems, a passivity towards other members of the species, a swamping out of predators, and a migration to bring the concentrations together. In two of these locations, the Platte and Fraser Delta, humans have altered habitats in ways that help. But, natural or otherwise, if any of the foregoing conditions do not exist, then a dispersed survival strategy, not concentration, would work to the species' advantage. For the vast majority of vertebrate species, that is the case. There is no special theory around the phenomenon of massed numbers. It operates as a fortuitous extreme of normal population processes. The responses to good habitat and food are no different for any species. The behavioural attributes that foster tolerance simply reflect a lack of intraspecies competition brought about by good living conditions, and the migration to accumulate in one place is no more mysterious than is any migration. Where all the requisite conditions are right, natural selection has smiled on these magnificent spectacles.

But the requisite conditions can go wrong, as explained in each of the forgoing instances. In fact, it is difficult to think of any species that masses together that is not in trouble. Flamingos, for example. We witnessed the hosts of flamingos on Lake Nakuru in central Africa, which is one of the world's greatest birding spectacles. The lake was set aside as a sanctuary for flamingos, which are totally dependent on its warm, alkaline waters. Their large aggregations provide safety, too, as they rhythmically feed while crowded against each other in massed, tight groups. Today the lake is compromised by toxic wastes from a copper industry. In addition, increased rainfall at times decreases Nakuru's salinity, and the microscopic blue-green algae that flamingos eat decline. Adding to the insecurity of the flamingos' future has been the introduction of tilapia fish, partly to control mosquitoes, that in turn has encouraged large flocks of competing white pelicans to occupy the lake.

Then there are the masses of screaming seabirds in places such as St. Lawrence Island in the Bering Sea, where large-bodied puffins nest in complicated burrows; black-and-white guillemots perch on bare rock ledges; tens of thousands of least and crested auklets seek nesting cavities among the rocks. Together, they make up the millions of pelagic seabirds that breed and reproduce on islands in the Bering Sea. Their numbers are endangered both by diminishing food abundance and, on some islands, by the presence of rats, cats, and numerous feral animals that have escaped from introduced fur farms on once predator-free islands. The nesting seabirds are easy targets for these predators.

Many other vast wildlife shows exist, such as the impressive Porcupine caribou migration that threads its way from the Yukon to calve en masse on the coastal plains of the Arctic Wildlife Refuge in Alaska. Although the herd is 200,000 strong, its sensitive calving grounds are threatened continuously by the demand for oil. Threatened, as well, are northern fur seals. On the Pribiloff Islands, thousands of seal heads bob in the ocean as they journey to their natal beaches to breed and give birth. For them, and the other marine-dependent creatures, a general shift in the biological makeup of the Bering Sea – especially the dangerous decline in the populations of nutrient-rich, oily species such as salmon and herring – threatens their future.

What about other spectacles that nobody will ever see again? Passenger pigeons once flocked in huge numbers over a vast terrain. It is incredible that they were stamped out within a fifty-year period, shot unremittingly in the millions. Despite adequate food and habitat, the species could not reproduce without the cohesion of large social groups, and flocks vanished as their numbers fell below the critical mass. Almost as dramatic was the great auk's demise. Leading the way to the species extermination was egg collecting and the clubbing and killing of millions of these flightless birds. Finally, the extensive feather trade for feather beds and quilts administered the death blow.

Is superabundance, once a time-tested opportunistic ecological strategy for some species, turning out, now, to be maladaptive? Is it a phenomenon in our world that today is under threat (except for alien, invasive species that wreak enormous damage)? Does evolution-proven safety in numbers now mean the opposite? Does the very ecology of these concentrations – their key set of environmental and biological variables – magnify the threat? Will our descendants see only museum dioramas of these spectacles, as we do today of passenger pigeons, great auks, and Carolina parakeets – species that, in their superabundance, seemed impervious to a danger that, even once it was realized by conservationists and wildlife managers, was unstoppable? The frightening answer to all these questions could be yes.

Today, conservation programs are directed towards "species at risk," a term defined by numerical rarity or rapid decline. Fuelling public concern and government action are international agreements like the Biodiversity Accord signed by most countries in the 1990s. What if another accord were signed that addresses biological phenomena at risk, directed specifically to ensure that no concentrating species declines in numbers past the environmental threshold beyond which massed superabundance cannot persist? Could we then save these wildlife spectacles of the world?

CHAPTER SEVEN

TEN THOUSAND REINDEER

No species can expand indefinitely in a finite space. It is inconceivable.

THE STARTING PLACE for thinking about population phenomena is the S-shaped curve, possibly the most meaningful symbol in biology. Take the capital letter "S," pinch each end with two fingers, and stretch it horizontally in opposite directions. What you get is a gentle curve. Now put axes around this curve – time on the horizontal axis and population numbers on the vertical axis – and you have a graph. That graph approximates the population growth curve of most things, plant or animal, virus or blue whale. The bottom left corner represents a population just starting out; the top right is a population's maximum size. At any time, a population has the potential to grow, and its growth would roughly follow the curve from the point where it currently lies to the top.

As always, there are exceptions, and a notable one is a population that erupts and then crashes. Instead of flattening off horizontally at the top of the curve, the line representing an erupting population continues to shoot straight up. The higher it goes, the steeper will be its crash, which is inevitable. In contrast, a population that reaches the top of the S-curve gradually either stays close to that point or declines more slowly.

The lower portion of the S-shaped curve approximates the biological potential of the species breeding in a perfect environment. As you run your eye up along it, the curve represents population growth picking up speed. But then, the environment begins to kick back, and on the top half of the curve, growth is slowed by what is called environmental resistance. It gets stronger as population density increases until finally all growth is stopped. This is the way it is for bacteria, ants, moose, deer, caribou, rabbits, fish, trees . . . and humans, unless the eruption-crash scenario becomes our destiny. Environmental resistance always does its job, causing every population curve either to flatten, decline slowly, or crash.

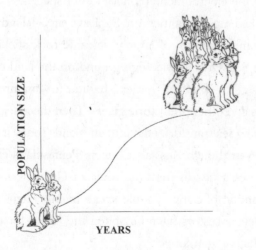

The top of the curve, where population growth ceases, represents what is termed the carrying capacity of the environment. Analogous to the number of groceries you can stuff in a shopping bag, or fish in a creel, carrying capacity is a practical, commodity-like idea that reflects the roots of population ecology in game management. Carrying capacity is simply the maximum density of a species that the environment can support.[1] The definition is simple, but the concept is slippery. Yet it cannot be ignored.

After many years of lying on the tundra, a stunted, weathered, lichen-covered antler, bleached by rain, snow, and sun, is displayed dramatically on a landing in our home. Spattered all over it are bright orange Xanthoria lichens. Touching it up, too, are black and grey splotches from several other lichen species, as if an artist has flicked paint-loaded brushes at it. It is indeed a work of art, the product of the incomparable, random hand of nature. We have kept it because it symbolizes so much: biological success and failure, human hope and defeat, and the tenacity of life that goes on despite the environmental foibles of our species.

We picked up the antler on St. Lawrence Island, a 4,600-square-kilometre chunk of largely volcanic rock sticking out of the Bering Sea. Once it was high ground on the land bridge that joined Asia to North America, habitat of mammoths and mastodons and other prehistoric giants. Then the glaciers melted, and the rising sea transformed it into an island. Today it is Alaskan territory, even though Russia's Chukchi Peninsula is closer, only 64 kilometres away. In the little town of Gambell, most of the island's handful of native people speak mainly Central Siberian Yupik. Their ancestors have inhabited the island for over 2,000 years, perhaps much longer.[2]

The day we found the antler was cold, with a strong sea wind whipping a pea-soup fog into our faces. We had flown to St. Lawrence Island from Nome, eager to see the boisterous masses of auklets, puffins, and other seabirds that inhabit the place. We had pitched our tent on the tundra behind town at the top of a cliff that curved out of sight to the north. One day, while stumbling along over wet hummocks, trying to make out the edge of the cliff that fell to crashing surf more than a hundred metres below, we spotted the antler. It was cradled in a muddy hollow among some sedge tussocks, partly concealed by moss. We pulled it out, examined it, and were struck by its symbolism. It looked like a caribou antler, but caribou didn't live there. Or maybe reindeer? The antler had history.

Once, reindeer were abundant on the island, a legacy of the innovative Presbyterian missionary Sheldon Jackson. He was a man only slightly more than five feet tall, driven by a determination not only to save the souls of native Alaskans but also to improve their plight here on Earth. Starvation was a constant threat, especially for the coastal and island people. Jackson devised a reindeer herding scheme that over a decade brought 1,280 animals from Siberia to Alaska.[3] Most were introduced on the Seward Peninsula, but some were also landed on offshore islands in the Bering Sea, and on July 23, 1900, the revenue cutter *Bear* brought twenty-nine reindeer to St. Lawrence Island to start up a herd there. A few weeks later, he sent forty-two more.[4]

Herding life was new to the hunter-gatherer Alaskan natives, and Jackson had difficulty finding anyone with any real interest in the new enterprise.[5] Fewer than three hundred people lived on the island, their numbers recently ravaged by starvation and disease. To help in the transition, he arranged for a family of experienced

Laplanders to accompany the St. Lawrence Island reindeer. He arranged, too, for a young medical student – missionary from California to manage the enterprise. In 1903, President Roosevelt helped by declaring the island a reindeer preserve.

After an initial reticence, the native people took well to the enterprise. A few reindeer were even domesticated and broken to harness; early photographs show them hitched up and drawing sleighs. Reindeer numbers rose slowly at first: 87 in 1901,[6] 212 by 1904.[7] In 1905 the population dipped to 189, suffering from what was diagnosed by the perceptive medical student as craziness: "In some cases the deer manifest symptoms akin to hydrophobia, fighting other deer, breaking up sleds, and even chasing the herders."[8] That year, 42 of 76 fawns born perished from the cold.[9]

Then the population took off. By 1917 it was estimated at more than 10,000.[10] This number, if true, represents one of the most dramatic ungulate eruptions in North American history. If you assume no mortality whatsoever, and that 50 percent of the herd were females, and that all females bred every year after one year of age and produced one young per year (the maximum for reindeer and caribou), in 1917 you come up with 9,342 animals. So, given some error around the estimate of 10,000, an eruption of such magnitude is theoretically possible, but barely.

A population growth rate like that represents what is termed biotic potential – breeding flat out – and with no mortality it is a scary thing. For example, if a female housefly lays on average 120 eggs, of which half develop into females, and there are seven generations per year, and if all survive that one year, then at the end of the year there would be 6 trillion, 182 billion, 442 million, 727 thousand, 3 hundred and 20 houseflies![11] A pair of robins lays on average 4 eggs, and the next year these 6 robins (3 pairs) end up as 18, then 54, then 122 . . . becoming 74,007 in ten years. In a 40-hectare wooded residential area with a starting density of one pair of robins per hectare – a reasonable density – after ten years the people would have to move out. They would be driven insane each morning by the singing of more than 3 million robins! If any animal population achieved and sustained its biotic potential for a few thousand years, even with normal lifespans, it would be expanding outward from the planet at the speed of light.

At 10,000 animals, the reindeer herd on St. Lawrence Island had exceeded any long-term, sustainable carrying capacity. But, thankfully, biotic potential is not all there is to population ecology, and finally environmental resistance, mainly in the form of starvation, kicked in. The herd crashed, leaving only a few hundred.[12] There it has remained for decades, living at a remote end of the island.

On nearby St. Matthew Island, a later introduction achieved an even greater density but ultimately fared even worse. The landscape on St. Matthew consists of the same cold tundra, but the island is much smaller, only 331 square kilometres. In 1944, 29 reindeer were released. By 1963 their numbers had exploded to 6,000, living at a high density of 18 per square kilometre (46.6 per square mile). One year later, the entire population had ceased to exist.[13]

A similar scenario played itself out on the nearby Pribiloff island of St. Paul. Four bucks and twenty-one does were landed there in 1911. They increased to more than two thousand by 1938, and then crashed. By 1950 only eight remained.[14]

Reindeer aren't the only creatures to have exhibited unstable populations on those cold, Bering Sea islands. So have humans. The human population on St. Lawrence Island was estimated at about 2,500 in the early 1800s by Russians, who then claimed the region. The inhabitants lived off the local resources – seabirds and marine mammals such as abundant Pacific walruses. However, by 1879 the human population had dropped by two-thirds to about 800, and by 1900 a mere 264 witnessed the coming of the reindeer. The populace remained low throughout the glory days of the reindeer, despite abundant food. Disease from increased outsider contact was a cause of continuing mortality. By the early 1950s the population stood at only 500, and by the mid 1980s about twice that, where it is today.[15]

In all these examples, ungulate and human, environmental resistance acted to halt population growth, exerting more and more pressure as the populations expanded until it crushed them. Environmental resistance makes the axiom true that no species can expand indefinitely in a finite space.

On the other side of the globe – New Zealand – and in a totally different environment – podocarp mountain forest – another introduced hoofed mammal, the red deer, demonstrated an eruption and collapse almost identical to that of the St. Lawrence Island reindeer.

The land mass of New Zealand slipped away from the supercontinent Pangaea and floated into the South Pacific before mammalian evolution had produced any widely distributed species of large mammals. Early Europeans who settled New Zealand bemoaned this shortcoming and decided to do something about it. They established acclimation societies, where success and prestige attended heroic efforts to introduce just about anything, plant or animal. Motivations were many: food, fur, trophy hunting, nostalgia. Government policies supported these introductions, providing protection while numbers built up.

In this way, European red deer were repeatedly brought to New Zealand: Stewart Island in 1902, Tongariro National Park in 1905,

Fiordland in 1901 and 1910.[16] They multiplied and spread over the country so rapidly that by 1922 their numbers had become a national concern. That year, the Forest Service made the first of successive submissions to parliament about the harm red deer were doing to forests and agricultural lands. Five years later, protection of red deer was lifted in state forests, and then in 1930 it was removed over the entire country. Still the red deer population expanded, by then well beyond its sustainable carrying capacity. In 1931, control operations began when paid hunters, called "cullers," were sent into the backcountry to shoot all the deer they could find. But the environmental devastation continued. In 1938, destructive floods and landslides of debris coming down deer-denuded mountains killed people and buried farms. Now the public was alarmed. The government beefed up culling operations in an all-out effort that continued for decades, but nothing was effective against the hoofed masses.[17]

Not until the early 1970s did humans gain control, when the scale of warfare against the red deer took on a dramatic added dimension. A lucrative European market had developed for venison, and, consequently, red deer farming came of age. The accompanying novel method of capturing red deer by helicopter finally brought the population to its knees. Deer were darted and tranquilized from the air, then slung beneath the aircraft and flown down to factory ships waiting offshore. In the early days of this aerial rodeo, a skilled pilot and gunner could take an amazing fifty or more red deer a day, and, despite high overhead costs, the practice was exceedingly lucrative.

By the early 1980s, when we were in New Zealand, the red deer population had collapsed. We hiked in all the national parks and had only one memorable red deer experience. While in

Fiordland one morning, following a twisting, high-country, mountainous trail, suddenly from behind, then from directly above us, came the roar of a helicopter flying at treetop level. Startled, we looked up to see the gunner leaning out of the door-less port side, pinioned in place by red strapping. In an instant the helicopter was out of sight. We heard the *thump, thump* of its rotor as it quartered somewhere ahead. Moments later the roar subsided.

That was our red deer experience. We never saw one in the wild in New Zealand, just some animals looking back at us despondently through high fences on red deer farms in the low country. By then the population had fallen so low that a pilot and gunner would be lucky to get one red deer a day, not fifty. With scarcity, however, the value per animal had escalated so that there was still a profit margin – a classic example of the economics of an unregulated resource that has spelled near-doom for many creatures: fur seals, sea otters, baleen whales, and bison.

The human populations of the South Pacific have exhibited instability, too. Easter Island provides one example, where the population before European contact was estimated at 15,000 but by 1864 had crashed to only 2,000.[18] Human populations collapsed on Mangareva Island and the Marquesas group of islands, the latter falling from about 90,000 at the beginning of the nineteenth century to only 5,246 when the French took their first census in 1887.[19]

How is it that the eruption-crash phenomenon has played itself out almost identically in reindeer, red deer, and human populations on opposite sides of the world? Without much difficulty, one could come up with many more examples: deer on Arizona's Kaibab plateau, koalas in western Victoria, Australia; wood bison in the Northwest Territories; the Anasazi people of Arizona and

New Mexico; the Mayans of Central America. It seems obvious that some general, universal processes are at work.

Yet eruption-and-crash is just one population scenario, and an exception at that, associated mostly with the introduction of new species. A different population trajectory is a cycle of growth and decline, following either a three- to four-year cycle in small mammals or a ten-year cycle in many species of northern wildlife. Populations increase for a few years to what seems to be the carrying capacity of the land, level off, and then decrease for a few years in a relatively smooth fashion, only to repeat the process over and over again. The rock ptarmigan we studied in Alaska had such a cycle (see Chapter 9). So did snowshoe hares, once so abundant in the southwest Yukon that their squashed bodies splattered the Alaska Highway. Entire mountainsides of dwarf birch turned brown from their destructive gnawing, even though dwarf birch is normally an unpalatable species rich in defensive alkaloids. The hares created a visible browse line on all the poplars at the height they could reach while standing on the snow; nothing capable of being browsed grew below that line. The next year, their numbers dropped, and they dropped again the next and the next until five years later you were lucky to see half a dozen in the entire summer. But over the next half decade, back they came to splatter the Alaska Highway once again.

So prevalent is wildlife's ten-year cycle in the northern half of the northern hemisphere that several books and a host of scientific papers have been written about it. Many species of grouse, ground squirrels, and fur-bearers show it to some degree.[20] Abundance followed by scarcity followed by abundance . . .

Various causes have been advanced. Most intriguing are the extraterrestrial ones, based on the observation that a disproportionate number of all the cycle peaks throughout the northern hemisphere have occurred sometime during the first five years of each decade.[21] Dominant among these theories is one that implicates sunspots, first suggested in 1942 by British biologist Charles Elton.[22] Sunspots are dark, cool spots on the sun with magnetic field strengths thousands of times stronger than the Earth's magnetic field. The number of sunspots on the sun's surface fluctuates with an eleven-year periodicity, and during sunspot peaks, the Earth receives considerably greater magnetic radiation. Sunspot periodicity, however, is slightly longer than the average cycle in wildlife, so that by the 1970s the cycles had fallen out of phase.

Then, in 1993, the topic was reborn with evidence that every several decades, when sunspot peaks were especially strong, they appeared to pull the cycles back into phase.[23] But this conclusion was challenged two years later when a lack of synchrony was noted between snowshoe hare population cycles in northern Canada and Finland.[24] If sunspots were causing wildlife cycles, they should be acting simultaneously everywhere. Anyhow, nobody has come up with a convincing explanation of how sunspots could accomplish this feat of inducing cycles. Correlations in biology without causal mechanisms to explain them represent weak evidence.

Nonetheless, the phenomenon of cycles, just like the phenomenon of eruption followed by collapse, argues for the operation of some kind of general population mechanism. Both phenomena have in common a cessation of growth followed by either a precipitous or a gentle decline, depending on how far the carrying capacity has been exceeded, then either recovery or extinction, depending on circumstances.

But cycles, like eruption and collapse, do not represent the normal population trajectory, either. More common is relative stability. While variations in numbers happen from year to year, they are limited to relatively small changes that take place in a seemingly haphazard way. Ecologists balk at the term "balance of nature" because that implies tight control. However, within broad limits, balance does occur. Neither extinctions nor rampant growth take place as often as would be predicted if changes in numbers were entirely random. Instead, most populations remain relatively the same year after year, or drop a bit and recover and fall back, exhibiting remarkable regulation at or near their sustainable carrying capacity.

For example, we studied a white-tailed deer population in Ontario's Point Pelee National Park that had been stable for twenty-five years. This stability existed despite abundant food, no hunting, no emigration, and no predation.[25] Similarly, a ruffed grouse population on our 2.5-square-kilometre (1-square-mile) study area in Algonquin Park displayed amazing stability for the decade that we studied it, even though the numbers of ruffed grouse living farther north typically fluctuated widely.[26] Population processes had a tight grip on both these populations, it turned out, through self-regulation (a topic explored in Chapter 9). Stable for a decade, too, was a moose population in Algonquin Park, for more immediate environmental reasons (Chapter 8).[27]

So it seems as though some built-in, universal resistance to dramatic population change usually operates in some manner, but then sometimes is modified and results in eruptions or cycles. Shouldn't we know what is happening, and why, if for no other reason than because somewhere in all this, human populations fit in?

———

Answers to population phenomena are fuzzy, unless we grapple further with the concept of carrying capacity. The definition – the maximum density of a species that an environment can support – is incomplete, and its shortcoming is especially notable when applied to humans. To illustrate, a professor in a Recreation Department was fond of handing out an assignment in which students were required to calculate the carrying capacity of a hypothetical campground of a certain size. In whatever way he worked out an answer, the question was misleading because campers could be jammed in until there was standing room only. An issue of quality is involved in carrying capacity, applicable to the recreation students as the maximum number of people experiencing a given level of environmental quality that the campground could support. For global or regional human populations, quality of existence is essential to any definition of carrying capacity. For wildlife, with fewer options and no technology, the quality of their existence is less of an issue.

Another twist on this all-important concept of carrying capacity is the dimension of time. Because in nature the carrying capacity of the environment often changes, a short-term carrying capacity is of less biological or practical significance than is a longer-term, average one. Humans provide an extreme example. In 1798, Thomas Malthus foresaw a halt to human population growth due to the imminent proximity of the carrying capacity, but since then, agricultural and technological advances have raised the capacity repeatedly, at the same time stretching the S-shaped curve vertically and putting off the imposition of a population ceiling. Similarly, wild populations experience altered carrying capacity from the effects of weather and other fickle environmental circumstances.

Yet another catch to carrying capacity, a contentious one, makes it one of the most information-laden ideas in ecology. It surrounds the contrasting notions of density dependence and density independence. The former holds that populations are regulated by environmental factors that exert more pressure as density increases. The latter holds that density is irrelevant. This debate began in the mid 1900s and is still with us today.

The principal protagonists in the early days were two British biologists, D. Lack and A.J. Nicholson, versus two Australian biologists, H.G. Andrewartha and L.C. Birch. Fuelling the debate was the claim made by Lack and Nicholson that density dependence is a logical necessity.[28] This claim amounted to a low blow, because it implied that the other camp did not even understand logic. To Lack and Nicholson, if populations are regulated, which all are, and regulation happens according to density, then it must happen *because* of density. In reply, the other side, the Andrewartha and Birch school of density independence, cited examples where weather suddenly staged an extreme event and killed in a population regardless of its size (or regardless of where it was situated on the S-shaped curve).[29] The debate seemed stalemated.

Out of this stalemate, however, has come a refinement that helps in envisioning population regulation at work. Picture an increasing deer population following the S-shaped curve and approaching its carrying capacity set by the availability of poplar, maple, and other important foods. As the population gets larger, it begins to eat up its larder. The animals themselves serve to reduce the carrying capacity of their own range. To depict that phenomenon, above the S-shaped curve you would draw a line curving down that represents that decreasing carrying capacity, eventually intersecting the S-shaped curve and stopping growth. In effect, the population itself "turns on" the environmental influence that eventually ends its increase.

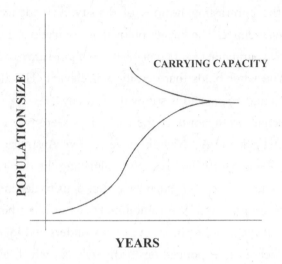

In contrast, density independent regulation occurs when the population is simply a victim of its environment, particularly weather, over which its numbers have no influence. Andrewartha and Birch may have been more attuned to this situation because they studied insects. Insects, being incapable of regulating body temperature, are more likely than vertebrates to be susceptible to the vagaries of weather.

But, frighteningly for humans, even weather-induced mortality may be density dependent. We may be the first species to make it so since massive populations of early bacteria infused the global atmosphere with toxic oxygen. Hurricanes and floods resulting from human-caused climate change must be viewed as a density-dependent effect, increasing in frequency with more of carbon dioxide – producing us. We are reducing our carrying capacity even as our population curve is still erupting up. In this self-defeating way we are hastening the point of our own population's demise.

———

Finally, carrying capacity has one more catch. Just as it has a time dimension, it also has a spatial dimension. Carrying capacity for what area?

Ecologists typically identify a study area for the practical reason of confining a field investigation to an area where you can get around. Our ruffed grouse study area in Ontario was 2.5 square kilometres (one square mile), seemingly small but deceptively large when traversing it on foot and in search of every grouse. In central Labrador our study area for caribou was 25,000 square kilometres (9,650 square miles); we covered it by helicopter. Both study areas, however, had one feature in common: they were each part of a habitat that was continuous in all directions. Choosing such a study area is common practice; it simplifies things. It removes the two additional variables, besides birth and death rates, that can influence population size: immigration (movement of animals in) and emigration (movement out). They are assumed to be equal.

Dismissing immigration and emigration in this way, however necessary for logistics, may introduce significant error. It assumes a uniformity of population distribution when, in fact, most animal populations are clumped – a dense pocket here, a sparse pocket there. That phenomenon happens because most habitats consist of clumped vegetation due to microclimate, or a patch mosaic after fire, or a change across a gradient up a mountainside or along a watercourse.

An even more important reason for uneven distribution is the metapopulation phenomenon, which is a term for the fact that most populations are really only subpopulations that, over large areas, exhibit ebb and flow movements and dispersal with others of the same species. Driving such movements may be so-called source-sink dynamics, in which individuals are forced through social pressures to emigrate from high-quality, or "source" habitats and move to low-quality, or "sink" ones. In source habitats, production exceeds mortality; in sink habitats, the reverse is true, and population persistence depends on immigration.

Such dynamics complicate population studies by making it difficult not only to define carrying capacity but to identify what may be truly one "population." Measuring immigration is especially difficult because distant individuals are not likely to be radio-collared or tagged and so cannot be recognized. Wildlife management is complicated by this difficulty. We were bothered when the Yukon government implemented wolf control in an attempt to increase the declining Aisihik caribou herd in the 1980s. A few years earlier, graduate student Dave Gauthier had been studying the nearby Burwash herd, which was living on mountainous tundra adjacent to Kluane National Park some eighty kilometres distant. Between two of his weekly aerial surveys, about one-third of the herd mysteriously vanished. They showed up in the Aisihik herd, however, where they spent the winter. A similar but reverse event, not wolf predation, could have been responsible for the Aisihik decline.

Similarly, over time, a Beaver Creek herd came to occupy an area adjacent to the Alaska Highway near the Alaska-Yukon border, where no caribou had been present for many years, and farther north a Steese-Fortymile herd occupied the rolling tundra near Eagle Creek, Alaska, for years, then vanished for years, then returned.

These examples leave the impression that all the various caribou herds in the southern Yukon and south-central Alaska, each given its own name and even hunting regulations to fit its perceived status, are really one big herd. Temporary subpopulations sweep and withdraw across the country, spinning off herds here and there, with animals periodically flowing among them. In time lapse, the effect is akin to pouring chocolate into a saucepan of milk and stirring it slowly but not to the extent of homogeneity.

It is mismanagement to consider the caribou herds individually, especially when deciding whether to implement a controversial intervention such as wolf control. Wildlife management still has some distance to go before it either adequately reflects our science or holds back on making prescriptions because of our ecological ignorance. Meanwhile, caution should prevail. Immigration, emigration – they are inherent and vital components of population processes.

The metapopulation phenomenon applies to humans, too. Leaders and concerned members of society know their nation's human immigration and emigration rates; setting policy for them is fodder for politicians and protest groups. These rates influence national and regional productivity, unemployment, poverty, social well-being, and, ultimately the most important of all, pressures on the environment. In the end, however, we are all linked together in a finite space, invariably subject to a common population fate.

Reindeer, red deer, passenger pigeons, white-tailed deer, moose, ruffed grouse, snowshoe hares: why has the accumulated knowledge about their population curves, eruptions, cycles, density dependence, density independence, and carrying capacity not informed us about our own global population and its hectic, irresponsible growth? Why does every politician, reflecting society in general, still see growth of all kinds as good?

The information from wildlife is not new. Back in 1933, pioneer ecologist Aldo Leopold wrote: "Man thinks of himself as not subject to any density limit. Industrialism, imperialism, and that whole array of behaviours associated with the 'bigger and better' ideology are direct ramifications of going to the limit of [human] potential. . . . But slums, wars, and depressions may be construed as ecological symptoms that our assumption about human density limits is unwarranted; that we may yet learn a lesson in sociology from the lowly bobwhite."[30]

CHAPTER EIGHT

LABRADOR WILD

All populations are embedded in environment. In seemingly clever, devious, generous, stingy, but really inadvertent ways, the environment organizes populations, sustains and buffers them, gives them room to prosper and grow. At the same time, it seems to track their welfare, measuring their needs against its larder. It keeps score. Then, in ways ranging from benign to brutal, it imposes its inescapable limits.

AT THE DISPOSAL OF THE ENVIRONMENT is a handy set of weapons to level populations, the most obvious being food supply, food quality, predation, parasites, disease, and weather. Because there are also alternative, more subtle methods, these obvious ones are grouped together by general agreement and labelled "extrinsic factors," which means that they lie outside the characteristics of the population itself. (Chapter 9 will explore the subtle "intrinsic factors" that work their craft from within.)

Among the set of extrinsic factors, if all else fails to curb population growth, food supply must do it – not absolute supply, but food packaged in whatever way the species finds palatable or available or otherwise acceptable.[1] Because climate conditions such as snowfall often reduce the availability of food (a fact well known but inadequately delineated in many studies), we coined a term for this point of ultimate population limitation: the "nutrient/climate ceiling."[2] So frequently do animals starve when food is still present but unavailable because they cannot get to it that the concept of a nutrient/climate ceiling has proven useful. For instance, mountain sheep commonly graze on traditional south-facing grassland slopes in winter. Some years they may consume all the available grasses and starve, although ungrazed grasses on another slope are within sight, because an intervening valley deep in snow prevents their travel. Similarly, woodland caribou can starve in years of low snowfall because they cannot reach up high enough to eat hanging lichens.

Food is a classic density-dependent factor, meaning that as any animal population increases it will eat up progressively more of its food supply until, finally, the increasing population curve hits the decreasing food supply curve and the animals starve. Examples of populations that have eaten themselves to death are easy to find, such as the introduced reindeer population on St. Matthew Island in Alaska (Chapter 7).[3]

Another classic example is the demise of the mule deer population that once lived on the Kaibab plateau north of Arizona's Grand Canyon. There, in that magnificent forest of native blue spruce, wildlife managers committed a costly mistake. Between 1906 and 1924 the predators – wolves and cougars – were culled. The result was an explosion of mule deer at an enormous rate of up to 20 percent per year. The deer herd increased from 4,000 to about 100,000 animals; then, in 1924, it collapsed. A forest super-visor wrote, "Deer died by the thousands and those that lived ate every twig till the whole country looked as though a swarm of locusts had swept through it."[4]

Lack of food has brought many other populations down, espe-cially those that exhibit typical boom-and-bust cycles. The numbers of snowy owls and rough-legged hawks in the Arctic decline peri-odically because the birds fail to breed in years when lemmings and mice are scarce.[5] Lack of winter food caused a crash of snowshoe hares in Alberta, documented in the late 1970s.[6] A great horned owl population fell in the Yukon in the 1990s when snowshoe hares declined.[7] The high Arctic island populations of caribou, sub-species *pearyi*, experienced severe icing of the vegetation between 1972 and 1974, which dropped the nutrient/climate ceiling, killing 90 percent of them.[8] Ever since then, Peary's caribou have been on Canada's endangered species list. Climate warming in the Arctic may increase the likelihood of similar icing in the future.

Conversely, food increases have resulted in population increases. Prairie chickens across the North American midwest expanded in both numbers and range after European settlement in response to the availability of grain.[9] Outbreaks of bark beetles in British Columbia from 2000 to 2007 resulted in increased numbers of three-toed woodpeckers and flickers.[10] In a controlled experiment, distribution of commercial rabbit chow in the Yukon's boreal forest – manna from heaven – caused an increase in what must have been very confused snowshoe hares.[11]

We thought that food was an obvious candidate for limiting the size of Labrador's Redwine Mountains caribou herd when we ini-tiated our study there in 1982. However, we found its population regulation was much less straightforward and much more intriguing than that. When we came, the land was still in winter lockdown. A high-pressure cell had sifted in from the west to press down like an unbearable load, stilling the snowshoe hares that normally dance on the moonlit snow, drilling through the fluffed feathers of the great horned owls, urging the ptarmigan to abandon the open slopes and huddle under the trees. Yet winter was losing its grip. The cold spells were not continuous any more. The sun radiated a warmth, discernable at least to the red squirrels, who left more tracks on runways among the spruces. The gray jays and red crossbills were nesting. The caribou were on the move.

Early that day we had left Goose Bay for a late-winter calving survey of the caribou herd with biologist Kent Brown. The Redwine herd ranges over some 25,000 square kilometres (9,650 square miles) of sprawling wilderness, a land of open taiga forests, lakes, rivers, and countless bogs – string bogs, raised bogs, every ecotype of bog. Punctuating the centre of this flat, featureless region are rolling tundra hills that look from the air like a rumpled blanket.

In late winter, the scattered caribou leave the surrounding lowlands and congregate in these hills. They stay only a few weeks. By the time the Labrador tea and the Kalmia bushes poke through the melting snow, they have dispersed across the lowlands again, the females to give birth in some clump of spruce or willow beside some lonely bog. In that way they play out an age-old ritual in an ancient and unmodified ecosystem shaped by time and honed by the harshness of the land.

Important to understanding that ritual was an appraisal of the caribou herd's production of calves. Caribou are easily visible at two hundred metres above the ground, but we needed to classify them by age and sex, count calves, and determine whether females with no calves trailing behind had a visible udder – evidence of lactation that would mean a calf had been born and died. To see an udder on a running caribou requires dipsy-doodle flying at treetop level. Sure, it was upsetting to the caribou, but only temporarily. It was more upsetting to me (John). I managed to hold out for the first few groups but then my stomach heaved, at first periodically, then continuously. Before long I didn't care about classifying caribou any more, but the worse I felt, the more caribou the others seemed to spot.

After a while we landed on a snow-covered pond, and the horizon took on some much appreciated stability. The previous year, Stu Luttich, the wildlife biologist for Labrador, had radio-collared a male black bear who was hibernating on a slope beside the pond. The bear's radio-collar was transmitting through the 1.4 metres (4.5 feet) of snow that entombed him. A small, partly encrusted breathing hole showed that he was still alive. He was an amazing bear, later aged by tooth sectioning at a ripe old twenty-nine years. The previous spring he had travelled 100 kilometres (62 miles) from his capture location in the mountains towards the coast then in late summer returned to hibernate close to where he had been caught. Not bothering to seek the shelter of a den, the bear had simply stretched out on the tundra and let the snow blanket him. But that was to be his last winter. When Kent returned after the spring melt, he found an emaciated carcass.

Off we flew again, and every sighting of caribou set off another in-helicopter ordeal. In tight confines, an airsick passenger is far from welcome. Airsickness tends to spread by association. Partially for my sake, but more for himself and the other occupants – his wife Joan, the pilot, and Mary – Kent agreed to eject me. Out of sympathy, or for safety, Mary got out too. In our packs was the normal survival gear – tent, sleeping bags, cookstove, food, although it would be some time before I had any interest in food.

For a while after the helicopter left, I slept. Mary walked in small circles around our snow-free knoll, the only bare ground in sight. Our snowshoes were in the helicopter, so she couldn't go anywhere else. Besides, in a few hours the helicopter would be back to pick us up – we hoped. The landscape looked the same in all directions; those were the days before GPS.

When I woke, we boiled some tea and discussed what we would do if the pilot failed to find us. With no options, we did not stay on that unnerving topic for long. Instead, we entertained ourselves by glassing our surroundings, looking for movement of any kind. Around us was a landscape where nature was the sole engineer. Very few people ever travelled into that expansive wilderness; flying over, we'd seen no snowmobile tracks down below, no hunting or fishing camps. In summer, the terrain was not traversable. Bogs, bogs, and more bogs. Stunted spruces, scattered birches, shrub heath, wet sedge meadows, and expansive fens choked with alkaline-loving dogwoods and cattails. Water everywhere. Mosquitoes everywhere.

As in most environments, the abundance of wildlife was inversely proportional to the presence of humans. Not only did an intact population of caribou live there, but also their dependents: wolves, black bears, and, to a lesser extent, lynx. Each year, more moose seemed to be moving in from the south, for unknown reasons. All the little creatures were present, too: pine martens, snowshoe hares, weasels, foxes; and the boreal birds: gray jays, crossbills, boreal owls. Here was an entire wilderness with all systems running as they should, counterpoint to urban blight, sinuous freeways, and brown, nitrous skies. As we scanned the wild scene around us, it was satisfying to know that such a complete functioning region still existed in this beat-up world.

Yet the peaceful beauty of that wilderness – was it about to end? The caribou survey we were conducting was part of a chilling, pre-development ritual – an environmental impact assessment. Eager engineers had acquired aerial photographs, conducted surveys, figured out the economics, sidled up to the right politicians at the right fundraising dinners – and they had plans. A second monstrous hydroelectric development had been proposed for the Churchill River to the south, with tentacle roads to northern Quebec. Already the first Churchill Falls dam, completed in 1970, had lain waste a huge tract of Labrador wilderness. We were to determine whether the Redwine Mountains caribou, which travelled as far south as the Churchill River, might be "impacted" by this engineering scheme – to put it cynically, whether they might object to spending part of the year underwater. We were required to propose "mitigative measures" – maybe snorkels? "Mitigative measures" is a placating term used by developers to gain government approval while often forging right ahead anyway.

Then there were the jets. Goose Bay airport is a NATO training ground for low-level jet flying – pilots learn how to fly undetected below radar in populated areas, drop a few missiles, and make it home in time for dinner. At the airport, the air traffic controller had held our helicopter back while a dozen menacing-looking, camouflage-patterned Phantom jets screamed down the runway ahead of us. Later, Kent corresponded with the base commanders from two countries whose pilots were training there to request that these flights not take place over the caribou calving grounds for three critical weeks in June. Likely he breached all kinds of international protocol to make unauthorized contact in that way, but one commander actually did respond and, amazingly, agreed to comply. The other ignored him.

On our patch of bare tundra, wrapped in wilderness, we were strangely disappointed when we heard the faint thrum of the helicopter returning to get us. Soon we would be whisked off on a twenty-minute flight to a place in the tundra hills that the caribou frequented, where Kent had chosen to camp. The helicopter would leave us again to the snow and the wind, but this time we would have camp chores to occupy us – tents to erect and dinner to prepare.

That evening, a visitor commanded our attention – a black bear, a seemingly hungry black bear. It would not go away. We had taken refuge from the wind and were cooking our meals in the doorways of our tents. The bear was interested and paraded back and forth in front of us, its level of aggression increasing. When it raised its hackles, I ran at it, shouting, feigning aggression before it could become more aggressive towards us. It responded by turning slowly and ambling downslope. Capitalizing on our minor advantage, Kent and I chased it, shot flares in its direction, and soon it disappeared in the haze of an approaching snowstorm. It was, however, not finished with us, nor we with black bears, because it turned out that they were an important part of the ecological system we were there to figure out.

The rain and snow hit just as we finished our meal, so we hunkered down in our respective two-person tents and there we stayed for thirty-six hours, stormbound. Any time we crawled out, the snow lashed our faces and the wind blew our breath away. We ate cold, dry meals in our tents, read some, slept a lot, and thought now and then about the helicopter we could summon by radio if things turned really bad. In 1903, the ill-fated expedition of Leonidas Hubbard and Dillon Wallace, who were trying to be the first white people to traverse the Labrador wilderness, had no such comforting backup. The book by Wallace describes many storms.[12] Before completing the trip, Hubbard died of starvation.

In the years before our study, the Redwine Mountains caribou herd had remained relatively steady at about eight hundred animals. However, population censuses of caribou provide only crude estimates at best. The problem is one of sampling. At a density in our study of roughly one caribou per 33 square kilometres, finding them was not easy, and the task was made more difficult by their patchy distribution. They spent the winters in scattered bands of a dozen to thirty or so animals, leaving vast areas with no caribou at all. It was impossible to fly over and examine all 25,000 square kilometres of their range, so instead, only a portion was sampled, specifically a set of fifty-four transects. Even that required flying several thousand kilometres.

Whatever the exact size of the herd, we could not help being impressed by all that empty landscape. Why were there not a lot more caribou? What was stopping population growth?

When a nutrient/climate ceiling has limited a population, it means that the population has avoided being limited at a lower number by some other extrinsic factor, such as predation or disease. Both predation and disease have the capacity to keep populations from realizing their full potential allowed by the availability of food.

Predation, like food, normally exerts its influence more strongly as the density of prey increases (density dependent). More prey triggers both an increase in the number of predators and the number of prey killed per predator – a dual effect. But predation is far more complex than that. The predator population often only tracks the increase in its prey population and does not cause its decline. Such a tracking situation happens if something besides the availability of prey limits the predator population, such as disease, human killing, or social intolerance within the predator population itself.

Even where it seems that predators are taking enough animals to stabilize or cause a decrease in a prey population, some more basic, underlying, predisposing factor may be at work, such as weakness of prey due to malnutrition. The many moose we examined that had been killed by wolves in Algonquin Park exhibited fat-depleted bone marrow, an indication of nutritional stress.

A susceptibility to predation may be even more convoluted. Woodland caribou in some places in western Canada may be limited by wolf predation,[13] but moose also appear to have played a contributing role. Moose have expanded their range in western Canada in recent decades, being absent in central and southern portions of British Columbia before the early 1920s.[14] Possibly, wolf densities have increased in response to the presence of moose, with a secondary effect being greater predation on caribou. The moose expansion, in turn, has been attributed to logging, which has converted much old-growth forest into young, early successional stages that grow more moose food.

Such a train of causation is of more than passing interest, because it should influence where wildlife managers apply remedial actions. Because woodland caribou are listed in Canada as a species at risk, government biologists find themselves under political pressure to halt further declines, and they have sometimes been guilty of invoking wolf control – that all-too-easy fix – instead of calling for curbs on the felling of the remaining old-growth forests and conversion to young stands that propagate moose. Such a misplaced focus on what may sometimes be only a prox- imate cause of caribou mortality – wolf predation – is, in the long run, doomed to failure. Caribou need suitable old-growth habitat and remoteness from humans, not the removal of predators.

In an attempt to assess predation on Labrador's Redwine caribou, Stu Luttich had radio-collared two wolves the previous year. One died a few months after being collared, but the other, a yearling male, provided data that showed he covered a home range of a whopping 3,250 square kilometres. Then, two months before our trip, this wolf died too. An autopsy showed that he had starved. While this information may have been indicative of a more widespread prey shortage for wolves, it was only one instance. Only when a statistically valid proportion of a wolf pop- ulation is radio-collared – somewhere around 10 percent – can you draw any conclusions about causes of mortality, another example of the constraints of sampling.

Thoughts about these sorts of mechanisms operating on caribou in the Redwine Mountains occupied our minds while we waited out the storm. They were not subjects of discussion, however, because the tents were flapping too wildly to hear anything below a holler. Thirty-six hours later we awoke to an eerie stillness and climbed out through a drift. We found nothing damaged, but some articles were missing – a six-pack of beer and one snowshoe. We had left both next to one of the tents, tucked under the edge of a fly. At first we were mystified, but after searching we found them on a knoll about a hundred metres away. Black bear revenge! The beer cans had been punctured and crushed by impatient teeth, and the gut stringing had been eaten out of the front and back of the snowshoe. As a calling card, or an act of derision, beside the snowshoe the bear had deposited a scat.

Fortunately, the snowshoe was not damaged beyond use. The melt was on and the tundra valley bottoms were filled with mini-lakes while slopes were still deep in snow. We had to wear both hip waders *and* snowshoes. We took the snowshoes off to wade through the deepest water but left the waders on full time – an energy-draining combination.

Kent had plotted out a course that led to another food cache he had set out earlier. En route, we frequently came across caribou tracks and found grazing sites where rocks covered in pulmonaria (lung-like) lichens protruded from the snow. Most remarkable were the excavations that the caribou had dug a full 1.25 metres (4 feet) down to expose the tundra. They were after Cetraria lichens, which, along with dwarf crowberry and blueberry shrubs, covered the ground almost everywhere except for the sedge flats in the valley bottoms. These snow craters represented a huge energy expenditure because some layers in the snowpack were hard and crusted.

En route, we frequently checked for the radio signals from any of the sixty-two caribou that Kent and Stu had collared earlier by aerial darting. These animals, mostly females, were vital to the study because they provided information on seasonal movements and, even more importantly, on causes of death. They also provided one of the more valuable clues to what might be limiting the population, from the blood that was drawn from a leg vein when they were captured. But that gets ahead of the story.

We were able to dial up only three radio signals, audible from hilltops, but as the animals were distant and in a direction away from our line of travel we did not pursue them. In the afternoon we made a lengthy detour along a meltwater stream and crossed, with some trepidation, at a place where the water swirled under the icy edge of a snowfield. Out on the snowfield we encountered a bull caribou picking his way along, head down, unaware of us. The sun glanced off the velvet of his new antler buds and etched a thin, ethereal line around his entire body.

Snow buntings in clean, black-and-white breeding plumage accompanied us much of the way, their tinkling songs reaching us in wind-whipped fragments. A rough-legged hawk came by to inspect us, tossing in the wind and then hovering in stationary mode, as they commonly do. Horned larks, newly arrived from the south, skittered away from patches of bare ground as we approached. Occasionally a willow ptarmigan with brown neck and white body, full nuptial plumage, would burst into the air and declare with guttural calls his right to a chunk of tundra.

Besides caribou tracks, we frequently came upon those of black bears and wolves, both species that were to figure in the explanation of caribou population regulation. Both the bear and wolf scats contained caribou hair. In addition, the bear scats were full of berries that had overwintered under the snow, their starches converted to sugars, making them extra palatable.

As the afternoon wore on, thin, milky clouds weakened the sun. The sky kaleidoscoped through pale yellow to coral, then velvety purple, and finally disappeared behind a bank of clouds. By the time we arrived at Kent's food cache, the land was brooding in lengthening shadow.

The cache was in a big metal box lying on the tundra beside a small lake edged with partly submerged, bluish ice. A welcoming party of five caribou were grazing on an exposed slope to the west.

The wind was strengthening as we set up our tents in the partial shelter of a low, rocky ridge. We cooked dinner in our tent doorways, bear-danger notwithstanding; shelter was insufficient anywhere else.

I got up once during the night, hoping to see the land flooded by moonlight, but clouds had engulfed the sky. Towards dawn the wind picked up even more – another storm was obviously on its way, after only a twenty-four-hour respite. An hour later, the wind was lashing heavy rain against our tents. There was nothing to do but hunker down again. The temperature dropped, and by noon the rain had changed to snow – a blizzard. Before it worsened, we dragged the food cache closer to our tents, where we could find it more easily. Then we lay there, watching the tent poles collapsing in the wind and snow. It was too dangerous to light our cookstoves. Years earlier, in such weather on Baffin Island, I'd lit a stove and the flame had touched the nylon tent wall. In a split second I'd found myself sitting on the open tundra, fully exposed.

Populations can be limited in other ways than by food and predation. Disease and parasitism can also do the job, and, as with food and predation, they commonly act in a density-dependent way: the denser the population, the more easily they are transmitted.

Wildlife literature contains many examples of populations limited by disease or parasites. Tularemia, a bacterial disease, decimated beaver populations throughout central Canada in the 1940s.[15] Sylvatic plague, a flea-transmitted bacterial disease that arrived in North America late in the nineteenth century, when piled on top of human persecution, crushed black-tailed prairie dog populations throughout central North America.[16] Rinderpest killed vast numbers of wild ungulates in Africa after its introduction via oxen during the military occupation of Ethiopia in the late 1800s.[17] The virus myxomatosis was famously effective in reducing the exploding populations of introduced European rabbits in Australia. Mountain bighorn sheep throughout their range in North America suffer periodic catastrophic die-offs from pneumonia caused by both a nematode (roundworm) and bacteria that infects their lungs and is transmitted by domestic sheep.[18]

Among birds, avian cholera is the worst infectious disease affecting North American waterfowl.[19] Acting together with avian botulism, it can silence a marsh; the two have done so repeatedly in places like Chesapeake Bay, where wintering waterfowl congregate. Recently, West Nile virus has cut an even broader swath in the number of species infected, but its role in reducing any specific populations is still unclear. Wildlife diseases are so important that a whole journal – *Journal of Wildlife Diseases* – is devoted to them.

And there are tragic examples in human history. In the fourteenth century, the Black Death (bubonic plague) in Europe killed about 20 million people over four years.[20] In just a single year, 1918, an influenza pandemic killed as many as 40 million people. Today we suffer from AIDS, malaria, and eruptions of avian and swine flu. Tomorrow? One of the most frightening prospects for humans is genetically derived drug resistance appearing in a wide range of potentially deadly pathogens,[21] the head-on collision of what are possibly the two most potent evolutionary forces on Earth: misapplied human intelligence, and the adaptive power of viruses and bacteria.

Yet disease as a general mechanism of population control is by no means straightforward. Evolution decrees that the ideal relationship between a disease or parasite and its host is accommodation. It is not adaptive for a disease or parasite to kill its host, as this leads to its own demise. Many examples can be found of a non-debilitating relationship when a host and its parasite or disease have had sufficient evolutionary time and exposure to adjust – physiologically, genetically, or epigenetically. For example, brainworm is a native North American nematode that kills moose, which are recent Asian immigrants, having arrived only some 9,500 years ago, but not white-tailed deer, which evolved in North America alongside it (Chapter 4).[22]

Also complicating the role of disease and parasitism is their ability to be carried by several taxonomically related host species that live in the same area. Then, a host species in low abundance may be victimized, in a sense, even driven to extinction because of the prevalence of another host species in its vicinity.

Because of these complications, ecology textbooks usually treat the role of disease and parasitism in an ambiguous way, with statements such as, "The matter of disease effecting regulatory controls on wildlife populations is a question stirring debate among biologists."[23] Dogging the issue even more is the perennial complication of separating proximate from ultimate causes, just as in predation. Extrinsic factors most commonly do not act alone but in concert. Even where disease can be shown as the immediate cause of a population decline, some other factor may have preconditioned its influence.

During our fourteen years studying wolf-prey dynamics in Algonquin Park, for instance, we commonly found moose that had died of hypothermia, but their deaths had been preconditioned by parasitism.[24] Winter ticks in the thousands parasitized individual moose, each tick becoming engorged to the size of a nickel. One effect of the ticks was hair loss, because these moose commonly rub up against trees and rocks to reduce the irritation. Another effect was anaemia caused by the ticks consuming up to a litre of blood. But the heavy tick infestation was, in turn, influenced by weather, particularly the absence of snow the previous spring, when the engorged ticks fell off moose onto the ground. When snow is present, many ticks perish and the cycle is broken. So, while the proximate cause of moose death was combined hypothermia and anaemia, the ultimate cause was winter ticks, and the "ultimate ultimate" cause was the previous spring's weather. Now, with humans influencing even the weather, an "ultimate ultimate ultimate" cause may be our climate-altering misbehaviour.

Unlike moose, caribou are reasonably free of parasites and disease. Warble flies that punch holes through caribou hides are a nuisance, and so are nose bot flies that lay eggs and produce larvae in their nostrils. However, in a review of principal causes of caribou mortality in the book *Big Game of North America*, author Tom Bergerud did not even mention either disease or parasitism.[25] In John Kelsall's *The Caribou* is a table listing six bacterial diseases and nine parasites that have been found in or on caribou, but also the comment that "deaths attributed to these phenomena were encountered only rarely."[26]

Nevertheless, caribou diseases may be out there waiting to strike. Caribou cannot live on ranges occupied by white-tailed deer, as an attempted introduction of caribou in Cape Breton Highlands National Park in Nova Scotia once demonstrated. Brainworm is lethal to caribou, but deer are not affected pathologically by them. However, through their feces, deer can infect snails, the intermediate host, which any caribou in the same area can pick up inadvertently while browsing.

As well, recent research has shown that caribou may have a genetic predisposition to chronic wasting disease.[27] This disease is caused by protein particles called prions which normally are non-pathogenic products of a specific gene carried by various species. However, the protein can cause devastating disease if its physical shape is changed by folding. That situation has occurred in mule deer, white-tailed deer, and elk, all of which transmit the particles from nose to nose in nasal droplets, or via feces. Research on prions is ongoing, and is especially interesting because prions may exhibit non-genetic inheritance (see epigenetics in Chapter 2) through the membrane of the egg. Prions, too, appear to be involved in several diseases that concern humans, notably "mad cow" disease, along with its variant, Creutzfeldt-Jakob disease, which can infect humans, and Alzheimer's. Fortunately for caribou, both brainworm and chronic wasting disease are not endemic in the north where they live, because deer and elk are not there. However, climate warming may stir the disease cauldron, and not only for these prion diseases.

Sometime during the night, the wind ripped our tent along the zipper of the door. We rigged up a tarp, but the blizzard made doing anything else impossible. We were not making any observations or collecting data. Eventually we resorted to using the single-sideband radio we had brought for an emergency. First we had to brave the storm and string its antenna across a gap between two stunted spruces. Then we called to ask if the storm extended down to Goose Bay and whether any helicopters could fly. A voice crackled back that if the storm let up a pilot would come for us. So we crawled back into our tents to wait. Back in 1903, Hubbard and Wallace could have benefited from a single-sideband radio. Today, satellite phones do the job even more efficiently. But for those intrepid explorers, isolation was complete – a human condition that has been banished by technology everywhere in the world.

The snow eased up somewhat while we waited, and we entertained ourselves by scanning the airwaves, unsuccessfully, for radio-collared caribou. A few hours later, the thumping noise of a helicopter became audible over the wind. We scrambled out of our tents and shot a couple of flares for the pilot to see. He landed on the ridge behind our tents, and we hastily broke camp. He had found his way by following the traceries of spruce trees lining the banks of rivers and streams that flowed off the tundra. For part of the return flight, too, the faint outline of trees was all that prevented complete whiteout.

When we neared Goose Bay the pilot radioed for permission to land, a routine procedure, but, disconcertingly, he reported to us through the headphones that permission was denied – the airport was closed due to the storm! Such a message from an airport control tower might be classed as an extreme case of following the rulebook, especially when no alternative airport was within fuel distance. We had to land somewhere, and the next safest place in the pilot's estimation was on the sea ice outside town. With no clear distinction between ice and sky, the pilot hovered the helicopter and very slowly edged downward until the struts contacted the frozen sea. There we sat out the storm.

Other than the caribou population census we'd conducted, the trip had yielded no tangible data. However, it had yielded insight into how the large mammal system was operating. Ideas and speculation often emerge in ecological studies when you are in the field. We pondered the late-winter snow conditions and wondered why the caribou concentrated on the tundra hills rather than staying spread throughout the taiga forest, as they had earlier in the winter. We thought about the black bear that drank our beer and the frequency of bear tracks we'd encountered in the hills, indicative of a possible seasonal concentration. And the wolf tracks. Were these observations related?

Kent was keen to answer these questions. Before coming to the University of Waterloo he had worked as a wildlife consultant conducting environmental impact assessments on caribou in the Yukon and Alberta. However, the terms of reference for environmental assessments are invariably superficial. All you typically are required to find out is what is happening, not why. It took Kent several years to figure out *why* for the Redwine herd.[28] The wait was productive, because eventually an elegant picture emerged.

The caribou remained spread out over this vast wilderness for as much of the year as possible. In that way, they managed to minimize the risk of predation. Each individual proceeded on its own or in a small group, avoiding any significant aggregation that might attract the attention of wolves. In small herds such as the Redwine, that strategy appeared to be effective. Later, helping support this conclusion were the results of our research on the next herd to the west, the Lac Joseph herd, with Guy St. Martin. That herd had once numbered about eight thousand. At that time it employed herding as a predator defence, both during and after calving, when the calves were tiny and vulnerable, and on migration, just as the huge herds do in the Arctic. However, numbers in the Lac Joseph herd fell, and when they dropped below two thousand, remarkably, their strategy switched to dispersion.

Dispersion worked fine for the Redwine Mountains herd all winter, even though they lived on an exceedingly deep snow pack, possibly deeper than encountered by any other North American caribou herd. Caribou have splayed hooves that reduce their foot load and manage to keep them on top of the snow, except when they dig for food. In late winter or early spring, however, the increasing warmth of the sun begins to soften the snow, making travel difficult. Compounding the problem is the sun crust, formed each night when the cooler temperatures refreeze the partially melted top layer. In these conditions, caribou stay on top of the snow pack for a few steps, then plunge through, then on top, then through – an exhausting situation.

To avoid it, the caribou, we discovered, migrated from wherever they were in the lowlands up to the more windswept tundra hills. There they exchanged their diet of "witches' hair" (Bryoria) lichens, dangling from tree limbs, for ground-dwelling Cetraria lichens that they dug for, and large, leafy pulmonaria lichens that grew on big boulders protruding from the snow. Food supply was no problem.

However, once the caribou were concentrated, even in relatively low numbers, they inadvertently created a set-up for predators. Black bears, normally not considered an ungulate predator, took particular advantage of this situation. Perhaps for them the presence of caribou was only fortuitous, since the tundra hills provided an attractive berry crop that the bears habitually exploited in autumn just before denning, and again in the spring. But on the tundra hills they also killed caribou, as was evident from the kill sites of several radio-collared caribou, mostly adult females. Bears emerging from hibernation need protein, especially lactating sows.

Wolves, it turned out, were equally important predators at that season. They no longer had to search so far and wide for prey. Consequently, predation by the two species of carnivores appeared to be the proximate limiting factor on the size of the caribou herd, exerting its strongest influence for only a few weeks. As soon as the snows melted sufficiently in the lowlands, the caribou left the tundra hills and dispersed widely once again, this time to calve.

Yet here was an interpretation trap, one that an unfortunate number of hunters, naturalists, and even biologists fall into. Our findings did not necessarily mean that if bears and wolves were somehow eliminated from this system, the caribou would increase. As counterintuitive as it seems, less predation sometimes may have no measurable effect on the number of prey. If the herbivore population is close to, or up against, a nutrient/climate ceiling to population growth, removing their predators will not result in a population increase because death from predation will just be replaced by death from starvation. So common is such interplay among the factors causing death that the phenomenon is called compensatory mortality: if one thing doesn't get them, something else will. Dodge a bullet and step on a land mine!

For the Redwine herd, that illusory effect of predation, in a modified way, seemed to apply. The critical evidence came from lab analysis of the blood drawn when the caribou were captured for radio-collaring. The levels of blood urea nitrogen turned out to be some of the lowest reported for their species. Urea is produced when protein breaks down, so the results meant that their diet was protein-deficient. We should have expected that. Protein is especially low in lichens, which constituted almost their entire winter diet.

Further evidence of nutritional stress was exhibited by low fat levels in the bone marrow of some of the caribou carcasses that were recovered after death by predation. If the marrow in the femur is gelatinous and red, rather than the off-white that indicates it is heavily impregnated with stored fat, it signals that the animal has used up this last stored energy reserve. However, we found no caribou that had outright starved, and the calf production of 80 percent of the adult females was near normal. So we concluded that the population was precariously near a critical nutritional threshold but was not quite across it.

And so, the elegant picture drawn out of the study was of a novel combination of extrinsic (environmental) circumstances limiting the size of the caribou herd: predation just before the calving period, preconditioned by a weather-induced concentration of caribou, but with food stress poised to replace the mortality caused by predation if, for any reason, predation declined. Relating this set of circumstances to the mechanisms of population limitation (described in Chapter 7), predation, as a classic density-dependent factor, was acting as a governor to smooth out the more extreme ups and downs of the caribou population that would have occurred as the result of the classic density-independent limitation by the weather. Because food resources were not depleted in the process, a reasonable stability in the caribou population resulted, rather than either the collapse, or decline and recovery, that typifies a cycle.

While this situation in the Redwine Mountains was novel, it is not novel for various environmental influences to work together to limit populations. Rather, it is normal, and there are many possibilities. For example, one explanation of the ten-year cycle in wildlife is called the consumer-resource hypothesis, where food limits population growth initially, but then predation works to steepen, deepen, and prolong a decline. Acting together but with time lags between them, the two may generate a cycle. This process, first proposed by mathematical ecologists, has gained support in some field studies. The dramatic snowshoe hare population cycle in the Yukon has been explained in this way based on experimental plots where food was added and where predators were removed.[29] Populations are being kicked around all the time by various combinations of environmental variables operating at varying strengths and shifting with changing conditions.

All this is good. Without the squeeze of these environmental assaults, and the fact that they ratchet up as populations grow, and the built-in redundancy of their actions, life on Earth would have become a disorganized mess and, like a supernova in space, exploded and blinked out long ago.

Today, more than two decades after our Redwine caribou research, the herd is still there. Other biologists have studied it, making use of our data to help interpret population trends. The trend is downward by 80 percent![30] Reasons are obscure but involve significant mortality of young females, which are under-represented in the demographic data.

And the wilderness, their home? The low-level jet training continues, although somewhat diminished by the termination of some of the foreign military agreements. In an apparent effort to recover such agreements, the Canadian military advertises on its website the merits of Goose Bay and surrounding region, with added surefire attractions such as a designated area to drop practice bombs, approval to fly as low as 100 feet, expanded clearance for supersonic flights in the training area, and opportunities for "ground exercises."[31] The land itself, its ecosystems and wildlife, have no value or significance in this expression of "Canadian military opportunity."

The additional dam for hydro power on the lower Churchill River, whose environmental impact assessment we were participating in, has not been built – yet. But just punch up "lower Churchill River hydro developments" on the web to see that whatever economic reasons prevented its construction in the past have not daunted its engineers, or seemingly the government of Newfoundland and Labrador. Ontario's premier has made supportive public statements, too, because transmission lines are planned to run there, as well as to New England.

Glad we have our memories.

CHAPTER NINE

ALL INDIVIDUALS ARE NOT CREATED EQUAL

If all that messed around with the destiny of wildlife populations were the direct push and pull of the environment (Chapter 8), then predicting their behaviour would be "easy." But there is more, another whole stratum to population limitation, one that flagrantly exposes various internal mechanisms of self-control. This additional stratum exists because all individuals are not created equal.

WHEN WE FOUND the third pika dead, we knew that our fledgling study was in trouble. The study had started out with such promise, because these diminutive members of the rabbit family had characteristics that made them excellent to study. They were easy to see, easy to catch, strongly territorial, and extremely aggressive to one another. They came in discrete subpopulations occupying small rock slides that were easily accessible.

Several weeks earlier, Jim Bendell, my (John) Ph.D. supervisor, had sent us off with a modest amount of cash to find a research problem. He had told us to go wherever we wanted and study something – anything – saying, "The species doesn't matter. It's the question that counts." The question that fixated him was one that continues at times to confound ecologists: "What regulates the size of animal populations?" By saying "study anything," Jim was asserting the belief that general or universal principles are out there awaiting discovery.

We drove around various parts of British Columbia and eventually pulled into a campground in Manning Provincial Park, which straddles the crest of the Coast Range. Behind the campground was a rock slide from which, every few minutes, came unfamiliar nasal beeps that turned out to be the calls of pikas. Active little animals, they perched on rocks across the slide, then scurried off in various directions, diving out of sight and re-emerging moments later to call again. Larger than a mouse, smaller than a rat, pikas appear rodent-like but are rabbits by virtue of their double buck teeth, one set of incisors situated directly behind the other, a peculiarity of all rabbits.

The question of what regulates population numbers concerned more biologists than Jim Bendell at the University of British Columbia. Motivated perhaps by books like *The Population Bomb*, by Paul Ehrlich,[1] and *The Limits to Growth*, by the Club of Rome,[2] the scientific establishment was looking to wild animal populations for insights into what increasingly was being recognized as an issue for humans. Knowledge on the subject had progressed far enough for the emergence of some lively general hypotheses that divided the Department of Zoology into ideological camps: a "food regulation camp"; a "predation regulation camp"; and an "intrinsic or quality-of-stock regulation camp." It was disconcerting for a newcomer trying to decide where to pitch a research tent. Each camp seemed convinced that it was focusing on an underlying and universal tenet of population regulation.

The greatest schism dividing biologists at UBC arose from the question of whether populations are limited by extrinsic, environmental factors (Chapter 8), or by some qualitative characteristics – behavioural or physiological – that are inherent in the populations themselves. Jim Bendell tended to be in the extrinsic, food regulation camp, having spent a sabbatical year at Oxford University where he was influenced by the internationally acclaimed proponent of the idea, David Lack. However, because of his own blue grouse research on Vancouver Island, he was intrigued by the ideas of the inherent (intrinsic regulation) camp. He had a sense, with mounting evidence, that blue grouse behaved differently on different clear-cuts, where forest succession was in different stages and populations at different densities. All this left me confused, but the glimpse given Mary and me by the pikas in Manning Park made intrinsic population regulation seem worth investigating.

On the rock slide in Manning Park, each aggressive male pika had staked out his own territory and gathered his own "hay pile." Pikas habitually assemble hay piles for winter survival, consisting of spirea or other shrubs situated nearby. The hay pile is normally stored in a large rock crevice somewhere near the centre of the male's territory, likely for security reasons. Pikas spend much of each day proclaiming their territorial rights. Like guard dogs, if another pika infringes on their territory they chase it away. Their life is one of near-constant stress.

Not ours, however. This was easy research. We sat in deck chairs below the rock slide, amassing our observations. Gradually we recognized the rock slide rules. The fellow who held the central territory also held the largest territory, which we were able to delineate from the position of other pikas when he gave chase. As well, he had the largest hay pile. Four other male pikas claimed parts of the slide, and all made surreptitious intrusions into each other's territory, which prompted quick retaliation if discovered. We were able to rank the pikas for aggressiveness based on each one's frequency of trespass and strength of counterattack. As well, we catalogued several other types of pika action: accumulation of goods; defence of possessions and property; chase and counter-chase, and attempted theft – all easy behaviours for humans to recognize. We did not see outright combat. These were civilized pikas.

Because one pika looks very much like another, we needed to catch and mark them. They were lured easily into a live-capture squirrel trap with a little spirea as bait. We went after the central territory-holder first, and after catching him, clipped a bit of fur off his left flank. To be sure that no other pika was using his territory, we left the trap in the same place that night.

The next morning, there was our marked pika – dead in the trap. We felt badly, of course, but that sensation was lessened the following day when a new pika perched on the same rock beside the big hay pile of the ill-fated owner. A neighbouring pika had been quick to move in.

Even more intriguing was the discovery that this new occupant was the pika we had ranked as second most aggressive on the slide, because his territory was now vacant. However, to be sure, we set out a trap for him. Next morning the same thing happened: dead pika. But now another new pika took over the territory, the third-ranking one. We put cotton wool in the trap to lessen any temperature shock and trapped him, but with the same unfortunate result.

Having messed up the pika population on the slide, but intrigued by events, we packed up and hiked to another slide a few kilometres away. There we ranked the pikas and set out our trap and again killed a pika. That was enough. While we had the beginnings of an interesting research project, you can't study the behaviour of dead subjects. Pikas, like other members of the rabbit family, are tense and high-strung and therefore susceptible to trap-stress. We returned to Vancouver and reported to Jim that we had successfully spent his money but had no specific study question and no study animal.

However, we had made our first foray into studying aggression and its possible role in influencing population size, a topic that would occupy us for years. Levels of aggression represent key differences among individuals in most socially evolved mammals and birds. Aggression has important physiological consequences that affect an individual's well-being, longevity, and biological success. Its manifestations, for humans, fill the TV evening news.

So when Jim announced that a colleague in Alaska, Bob Weeden, wanted a student to study social behaviour in rock ptarmigan, we were eager to go. Bob had collected a decade of population data, banding ptarmigan on the big, rolling tundra hills east of Fairbanks. He had recorded the dramatic cyclic changes in numbers that typify this bird. However, he could identify no corresponding cycling environmental variable: not weather, predation, or disease.

Bob, like everyone studying population ecology, was aware of the various theories concerning changes in the quality of the stock – the intrinsic factors – that might play a role in causing the cycles of abundance exhibited by many arctic and boreal birds and mammals. Cyclic species have received particular attention because, unlike steady populations, they provide the advantage of a template of annual changes in numbers that can be measured against various possible environmental causes.

Dominant among the intrinsic theories was the Chitty Hypothesis, a lively topic in Vancouver because Dennis Chitty was a faculty member at the University of British Columbia. He stalked the halls or sat in lecture rooms, looking at neophytes like me through thick-rimmed glasses and evoking unfounded fear in those in the other population regulation camps.

Chitty proposed that cycles are caused by different selection pressures on populations at high and low densities.[3] When populations are high, competition among individuals dominates and the most aggressive are the winners. These aggressive individuals carve out large territories because they are capable of defending them, but in so doing create a doomed surplus of less aggressive individuals that are kicked off the prime habitat. As a consequence, the population begins to fall. Adding to the decline, mortality increases because these highly aggressive individuals have not been selected to withstand other stresses such as weather, that is, they exhibit low viability. At a lower density, being aggressive to compete with others is not as important any more, and so numbers rise as more tolerant and more viable individuals increase in the population, until, at high density, aggression becomes most important again. Chitty called this hypothesis "balanced polymorphism," meaning a balance between two genotypes in the population that are differentially selected at various stages in a cycle.

Unlike many scientists, Chitty spent his career trying to disprove his own hypothesis, not prove it, because proof is always tentative, whereas disproof is final. He did not disprove his hypothesis, and wrote a sorrowful book about his decades of research called *Do Lemmings Commit Suicide?* [4] in which he reflected on his failure to prove himself wrong.

Important in Chitty's Hypothesis is the role of territorial behaviour, here resulting in limits on population growth by denying land to some individuals (the doomed surplus). This role is considerably different from territorial behaviour that merely spaces animals out to fit the available land or resources, passively influencing the density of the population by acting "like a supple, expandable disk, increasing or decreasing with the availability of food" (as described in Chapter 5). In contrast, here territorial behaviour is more like a compressible rubber ball held in the hand. A territory-holder may initially reduce the size of his territory to accommodate a neighbour, but as more potential neighbours come, he becomes less and less willing to do so, eventually refusing. Then the doomed surplus is created, and in that way the population limit is set.

This essentially self-regulating role for territorial behaviour is not easy to document for logistic reasons. Not only would you have to show that a surplus population of non-breeders exists, but you'd need evidence that their failure to breed is being caused by the breeders, and if allowed to breed they would do so successfully.[5] To collect data or manipulate any wild population to show all this in one breeding season is difficult. Consequently, evidence of territoriality playing this population-regulation role has been documented thoroughly in only a few species, such as Scottish red grouse, Australian magpie, Alaskan Cassin's auklet, and Vancouver Island song sparrow.[6] Likely, population regulation through territorial behaviour is far more common, but it is difficult to know; sometimes nature puts up an effective smokescreen.

A competing theory, called the Christian and Davis Hypothesis, was put forward by two California biologists. It deals with the physiological rather than genetic consequences of aggression.[7] This theory agrees that aggressiveness is a desired trait in a dense population, but argues that it comes with severe physiological costs. Aggressive situations stimulate a series of hormonal changes in mammals and birds. These changes prepare the body for the aggressive situation, more accurately the "fight or flight" encounter, because aggression is most commonly and accurately called "agonistic behaviour," with a component of fear attached as well.

To explain it, picture yourself being rear-ended while driving to work. You jump out of your car, ready to slug the *!@#* in the car behind. But your body needs to prepare you. Your anterior pituitary gland quickly pumps a hormone known as ACTH (adrenocorticotropic hormone) into your blood. In seconds the ACTH reaches your adrenal glands, situated just above your kidneys, and stimulates them to produce several of its own hormones, including adrenalin. As the adrenalin courses through you, it causes a set of physiological changes. Your heart and respiration rates increase, your pupils dilate (you want to see the subject well), your peripheral blood vessels expand (you want lots of blood in your arms so you can really slug him). Your adrenals also produce glucocorticoids, hormones that promote cell division, which may help repair your wounds later if the subject happens to fight back.

All these hormonal changes are adaptive, meant to help you handle stress. However, if an animal is subjected to repeated or constant stress – as it may be in a stressful population situation, especially if it is a social underdog – eventually the "general adaptation syndrome," as these hormonal changes are called, just packs it in and stops working. Then an individual is left without the necessary internal adjustments, and its survival is at risk. Or, in less severe circumstances, its reproductive output drops. In either case, in a stressed population numbers will fall until the population density is so low that aggression is no longer a useful trait. Then the population will build back up until aggression kicks in at the top of the cycle again.

A final theory was put forward by the beleaguered biologist V.C. Wynne-Edwards, who also took into account the quality of the stock. Wynne-Edwards observed that many species exhibit mass displays that have no obvious function. For example, in autumn, starlings commonly congregate in treed, urban areas. Every so often, without provocation, they all fly up, circle a few times while squawking loudly, then land again in the same trees. Wynne-Edwards wrote a thick book called *Animal Dispersion in Relation to Social Behaviour*[8] in which he documented a wide variety of similar mass displays and proposed a function for them. It was, he believed, a form of unconscious self-census in which the birds get an impression of the population's size. Some individuals, presumably the physically or socially inferior, feel the tension of being in a particularly large population most acutely and simply leave, going instead to some sub-marginal habitat where they are disadvantaged in some environmental way. The result is adjustment of the population size to the available resources.

Wynne-Edwards was attacked vehemently by the more aggressive of his fellow biologists because his hypothesis required some level of altruistic or self-sacrificial behaviour. Self-sacrifice flies in the face of natural selection, which holds that individuals are out for their own good. He eventually rescinded his hypothesis, but not before the subject of altruistic behaviour received heavy scrutiny. Out of this scrutiny came an understanding that natural selection will indeed favour self-sacrifice when it is made for genetic kin. For instance, two offspring have the same number of your genes in total as you do. Four grandchildren the same. This mathematics can be extended to the requisite number of first cousins, etc. In social groups that include an individual's own genes living in others, self-sacrifice is a winning strategy if the number of your genes saved exceeds the number you have in you.[9]

The spring following our pika debacle, armed with this set of theories, off I went in a rented pickup to Alaska, where Mary joined me when her teaching job ended. Snow still packed the tundra valley bottoms and corniced the ridges. Spicing the eternal wind was the tinkling of Lapland longspurs and the "tulik" of golden plovers. Along with us came two pointer dogs, one a pup named Tulik that lived with us for sixteen years and another with the less innovative name of Maggie. We rented a cabin just below the treeline near the gravel road known locally as the Steese "Highway" (an anachronistic name for a narrow, winding dirt road). There, gray jays poked around the outbuildings caching their epicurean tidbits for the winter, and grayling almost leapt into our frying pan from the little stream that ran by the door.

Up on the tundra, the male rock ptarmigan were spaced out on their territories, proclaiming ownership with dramatic flights and guttural on-the-wing calls. They were still in all-white winter plumage, with a blood-red comb over their eyes. The females had moulted to brown a little too early this year; on the snow they were obvious to gyrfalcons that would stoop out of nowhere and pluck them off.

We had worked out a solid research design: Jim and Bob both liked it, and together they found the necessary funding to keep us alive and put it into operation. Included in the budget was enough money to hire two assistants, one of whom was from Florida and had never spent a day in the wilds. He did most things wrong, including putting kerosene in a Coleman lamp that almost blew up the cabin, and burning his eyebrows off by dousing our garbage with Coleman fuel in a 200-litre (55-gallon) drum, then watching closely what would happen when he dropped in a match.

Our design called for both field and laboratory components. The latter involved raising chicks from eggs collected on the tundra and brought to an aviary at the University of Alaska's Institute of Arctic Biology in Fairbanks. We maintained a constant aviary environment for the three summers that the experiment lasted, with artificial lights on a strict regime, standardized food (chopped lettuce and eggs), standard pens, and a flock of domestic chicks as a control. Great care was taken with the eggs, which we transported to the aviary in thermoses filled with heated grain to maintain a constant temperature for the three-hour drive. At the aviary we placed the eggs in an incubator, then, when they hatched, put each brood in a separate pen. When the chicks were less than one hour old, we presented them with a mirror, repeating this procedure every three days all summer and again the following spring.

The results were dramatic. Each chick displayed one of three types of behaviour consistently from day one, demonstrating that the behaviour was hard-wired, that is, inherited. One behavioural morph showed curiosity, standing before the mirror, scratching and pecking gently at it, and giving thin, high-pitched, *seepy* sounds. Another behavioural morph attacked, running and striking hard at the mirror, sometimes losing its balance and collapsing in a heap, then getting up and doing it again and again. The third morph would take one look at itself in the mirror, retreat to the far end of the cage, and forever after stay away.

On the tundra, we caught and marked chicks with tiny metal

clips, measured body characteristics, recorded their escape distance, and ranked their physical assertiveness while in the hand. Maggie soon taught Tulik the skills of being a bird dog. When we flushed a brood, the two dogs would scout around, nose to the tundra, and suddenly freeze, tail rigid, with a motionless chick about ten centimetres away. There, both dog and bird would remain until we arrived. During our processing, the dogs would scout around and freeze on another chick. In that way we collected data on entire broods.

Over the three years, as the wild population peaked and was dropping, the proportion of more aggressive, or agonistic, individuals increased: from 12 percent to 28 percent to 49 percent, just as both the Chitty and the Christian and Davis hypotheses predicted. Here was convincing evidence of inherent changes in behaviour taking place at a population level.[10] Here, as well, was rapid evolution, called "sequential" in cases like this because it involved a fluctuating rather than permanent change in the proportion of genes, as occurs in the formation of new species.

Bolstering this direct evidence for changes in the quality of the stock that took place over the three years was supporting evidence from our analysis of Bob's ten years of population data. It showed a striking parallel trend in all the phases of the ptarmigans' yearly biology. When the population was increasing, so did the clutch size per hen, the percentage of eggs that hatched, the survival of both young and old chicks, and the over-winter survival of juveniles. When the population was declining, everything reversed. This result could not be attributed either to chance or to environmental contingencies.[11]

To distinguish between the two hypotheses supported by our results, at the end of the last summer we sacrificed some of the captive population and delicately extracted and weighed the adrenal glands. The Christian and Davis physiological stress hypothesis predicted a heightened adrenal activity. If correct, we would expect to see larger than normal adrenal glands. We didn't. The weights of the glands in the three behavioural morphs did not vary, which meant our data supported the Chitty hypothesis by default.

Left unresolved in our study, however, was whether genetics was involved in the differences in the inherited levels of aggression, as Chitty proposed. Two other options for inheritance are possible. One involves the nutritional quality of the albumin or egg white, which reflects the nutrition of the hen just before or during egg-laying. The effect of maternal nutrition on offspring is well known, having been studied diligently in humans. As well, many studies have indicated links between nutrition and aggression, in livestock, humans, and occasionally in wildlife.[12] One study showed that territorial aggression in Scottish red grouse (same genus as rock ptarmigan) could be traced back to maternal nutrition, which affected the quality of the egg, but no specific nutritional element was identified.[13]

A nutritional contender, however, that is receiving considerable attention in humans is the essential amino acid tryptophan, which animals cannot synthesize and must ingest from plants or microorganisms. Tryptophan is necessary for the body to make serotonin, a neurotransmitter that functions in a variety of ways. Its shortage is known to increase levels of anxiety and aggression.[14] The synthesis of trpytophan is biochemically complex, requiring dietary phosphorus and other elements.[15]

Nobody really knows if tryptophan or any other nutrient plays a role in wildlife cycles. To be a candidate it would have to cycle, maybe driven by some extraterrestrial influence like sunspots. At present, little evidence links sunspot activity with plant nutrition beyond one study of ultraviolet light. In that study, the train of consequences is as follows: low sunspot activity leads to a thin ozone layer, which results in high levels of surface ultraviolet, which stimulates plants to produce protective pigments but at the cost of reduced alkaloids and other chemicals that protect the plants against herbivores.[16] In theory, then, herbivore diets should be better at the low phase of the sunspot cycle. The populations of several forest moth species have responded accordingly. However, this discovery is still a long way from providing a universal explanation. Summing up, a recent review of nutrition related to human aggression, which could be extended to other animals, concluded that "the literature offers numerous clues, but little scientific verification."[17]

The second and more substantial option for non-genetic inheritance, and hence for our ptarmigan results, was not even suspected back when we did our research. It is epigenetics, the action or inaction of specific genes caused by changes in the environment (see Chapter 2). In this case, the genotype of ptarmigan at both high and low population densities may be exactly the same, but gene expression may differ due to differing selection pressures. This explanation, today, seems more likely than Chitty's one that envisions different genotypes at the highs and lows of cycles, because epigenetic inheritance can alter and adapt to rapid environmental change far more quickly than genes can be selected.[18] For that reason, natural selection would favour the epigenetic route whereby in just four or five years a population becomes dominated by newly adapted individuals.

The ptarmigan population still cycles in Alaska, and the ability to sort out genetic, nutritional, or epigenetic explanations has been greatly enhanced in recent years. Because aggression in humans is a hot topic with potential therapeutic application, here is a research set-up just waiting for the right biologist with the right background to exploit.

Each spring, when the hazel shrubs string out their tiny scarlet blooms on leafless branches and the winter wrens weave through their musical scores, then the male ruffed grouse return to their drumming logs. There, for four or more hours each morning over a period of three months, they mount a Herculean effort to attract a mate. Each drumming sequence lasts about ten seconds, during which the bird's wings beat the air with increasing speed until they are just a blur. Drumming at an average of about eighty times per day, multiplied by some ninety days, adds up to more than seven thousand sequences a season, a considerable energy demand. Yet the incentive is great, because at any time a female may wander nearby and be attracted. Or so the general understanding goes.

In a 2.5-square-kilometre (1-square-mile) block of Algonquin forest, for each of the ten years of our study, between nineteen and twenty-two drumming males enacted this genetically programmed ritual. Each bird occupied an "activity centre," which is a form of territory where boundaries are less important than a central core. Probably an equal number of females mated and produced an average of somewhere between five and ten chicks, increasing the population accordingly. Yet each spring the number of drumming males remained roughly the same. Years ago, ruffed grouse in Ontario exhibited a classic ten-year cycle of abundance,[19] just like rock ptarmigan in Alaska, but since the 1950s, for unknown reasons, the cycle has become irregular. Most regional biologists now refer to ruffed grouse fluctuations, not cycles. However, the population in Algonquin Park was stable.

The most obvious explanation was that some environmental bottleneck was at work. Yet two very different forest types were represented about equally on our study area. Sandplain covered the southern half, which was so flat that, on a cloudy day without a compass, south had the uncanny ability of becoming north. Growing there were pines, aspen, spruces, and firs, which replaced a dominance of pure aspen that followed a fire sixty-three years earlier. In contrast, the northern half of the study area consisted of rolling hills of deep till soils that grew sugar maples and yellow birch in the uplands and spruces and firs around beaver ponds and bogs. Compounding these differences even more, two years after our study began, spruce budworm attacked. On the southern sector, most of the spruces and firs died, and a maze of poplar seedlings and raspberry vines progressively pushed its way through the limbs of the fallen trees. On the northern sector, budworm impact was confined to the lowland conifers alone. Given such different environments, it was difficult to see a common constraint on the population of ruffed grouse.

Yet the number of grouse pumped out by both sectors remained roughly constant. It was always twice as great on the sandplain, both before and after the budworm attack.[20] Ruffed grouse have a close relationship in winter with aspen, whose buds they eat, and aspen was always much more common in the south. For aspen to affect population size, however, it must influence either birth rate, death rate, or dispersal. We discounted birth rate because each autumn young grouse invariably undertake a "fall shuffle," moving a few kilometres away from where they are hatched. We discounted death rate, too, because it was identical on both halves of the study area. That left different rates of dispersal, a difficult topic to study, but possibly gauged through measuring the aggression that was expected to precede and cause dispersal, according to both Wynne-Edwards's and Chitty's theories.

So, we set about measuring aggression in the two subpopulations, once again using mirrors, this time set up on drumming logs. Again, the response was dramatic. A male, upon arriving at his log at first light and hopping up to his usual drumming perch, catches a glimpse of his image in the mirror. "A female!" he thinks (both sexes have identical plumage). Within seconds he transforms into full display: head erect, iridescent black neck-feathers puffed out, tail fully fanned and raised vertically to form a backdrop for his stately pose. He struts around the mirror for about one minute, occasionally glancing obliquely towards it. Suddenly he pauses and stares at the image. His brain grasps that something is wrong. "How can I have been so stupid?" he thinks. "It isn't a female – it's a rival male!" Immediately a whole different set of programmed neurons fire. Within twenty seconds he is reconfigured: head down, body feathers sleeked, tail folded and projecting straight back. He is torn by conflicting impulses to attack or flee, the typical agonistic "fight-or-flight" dilemma.

Some males resolved the dilemma within seconds and attacked, half running, half flying at the mirror and striking it with their beak and shoulder. They would fall off the log, get up, and attack again. Other males slipped into exploratory behaviour, gently pecking at the mirror for hours. Others pecked a few times and left. Here were exactly the same three behavioural morphs that we had found in ptarmigan, and here, as there, likely reflecting inherited levels of aggression. We expected to find different proportions of the three morphs on the two halves of the study area, responsible for their different densities, but surprisingly there was no difference.

So we tried another approach to get at a better understanding of aggressive behaviour, this one focused on occupancy of activity centres. Did aggressive grouse set a bottleneck on the number of breeders by denying activity centres to less aggressive males? To test this possibility we live-captured various grouse on their drumming logs with a mirror trap and transported them off the study area. The birds had to be moved more than three kilometres or they would be back on their usual log the next morning, even when transported in a paper bag over a convoluted route by canoe, truck, and on foot.

In two-thirds of the removal experiments, the log remained vacant. In the remaining one-third, however, within twenty-four hours a new grouse, previously a non-drumming male, was drumming on the log. Apparently these replacement males were being denied drumming logs by the previous proprietors. Yet confusing this interpretation was the fact that every year, vacant activity centres were available that these replacement birds could have used. Over the years, the grouse on our study area used more than fifty distinct activity centres, only half of which were claimed in any one year. In some instances, vacant activity centres lay right next to ones where we removed a male and got a replacement.

The only explanation for this unexpected observation is that some grouse attach themselves to breeding males and are psychologically influenced to stay near them rather than establish their own activity centres. Perhaps they are the doomed surplus that theory predicts, enslaved to wait their turn, if it ever comes, or die without leaving genes to the future, if it doesn't.

Alternatively, however, these non-drumming males may be the clever ones. Perhaps they intercept a female attracted by a drumming male and breed without having to take the risk of advertising from a drumming log, which sometimes attracts predators. Such a strategy may work for some males, but if it predominated, then drumming would be a serious handicap to breeding, and without a positive selection pressure it would be eliminated. Most likely these replacement birds were indeed surplus, made so by the aggressiveness of the territorial birds that were setting a population limit.

Why that limit was different on the two halves of the study area, however, was not explained by this discovery of socially mediated occupancy of drumming logs. One possible explanation, but there was no way to study it, is simple habitat selection. Grouse, like most species, have an innate ability to recognize suitable habitat, maybe even to rank it.[21] For instance, no grouse live in the urban core of large cities, not because birth, death, or dispersal rates handicap a population there, but because grouse do not choose to settle there. Fewer grouse may have decided to settle in the northern sector of our study area because they detected some habitat deficiency, most likely less aspen.

Here, Jim Bendell enters the picture again. He eventually abandoned his study of blue grouse on Vancouver Island and moved to Ontario, where he began studying spruce grouse, again saying "the species doesn't matter." Several populations of spruce grouse in central Ontario showed remarkable stability at different densities, just like our Algonquin ruffed grouse. After tracking the characteristics of spruce grouse over seven years in their favoured jack pine forests, he concluded that females, rather than the males, were determining the spacing and thereby setting the density. They did so according to their perception of food availability to rear their young, which varied in different-aged forests.[22] The males just went along, adjusting to the availability of females. He explained that, "Because it is easier to study displaying males in most species of birds, considerable research may have focused on the wrong sex."

And so, we step back to highlight the subject of population regulation, which has been, and still is, such an important focus in ecology. Its earliest expression can be traced back to 1798 when Thomas Malthus speculated on the grim future awaiting humans. Since then, various embellishments have been added, like decorations on a Christmas tree:

- 1828: the S-shaped curve of population growth,[23] largely ignored until biologists saw its value in making predictions in the 1920s (Chapter 6);
- 1920s: the spacing role of territorial behaviour (Chapter 5);
- 1930s: the concepts of density dependence/independence, showing that most mortality factors increase in severity with density and invariably end population growth (Chapter 6);[24]
- 1940s: the idea of compensations between mortality factors,[25] that is, "if one thing doesn't get them something else will" (Chapter 8);
- 1950s: the separation of extrinsic factors (direct environmental) from intrinsic factors (this chapter);
- 1960s: the recognition that environmental stress can work subtly on animal physiology, behaviour, and gene selection, and an understanding of a difference between proximate and ultimate factors influencing populations (this chapter).

Since the end of the 1960s, except for ideas related to "consumer-resource dynamics," which expanded in the 1970s and more so in the 1990s (Chapter 8), new overarching concepts to explain population regulation have not been appearing. New analytical approaches do keep showing up, though, the result of computer applications. These include the mathematical wizardry that allows the allocation of relative importance to various environmental variables, and programs like the "population and habitat viability analysis," which has become a mainstream technique for identifying ecological problems faced by endangered species. Such techniques sometimes spin out important management prescriptions, especially for endangered species, pests, park management, environmental impact mitigation, and other applied fields. But the depth of analysis facilitated by computers today has not brought any new general theories to light.

This observation raises two interesting but conflicting assessments of the state of affairs in understanding population behaviour. One is that to see things more clearly, more studies are needed. The complexity of populations, interlocked as they are in ecosystems, may be so great that we require more fodder before any new overarching principles will become apparent. Science progresses first through the accumulation of studies, then the appearance of some deductive thinker who can take those studies and find common trends, patterns, and explanations.

Alternatively, perhaps there is no underlying universal mechanism, a "one size fits all" explanation for how population regulation occurs. That may be the real message of all the research over the past forty years, in the period since new ideas dried up. Charlie Krebs, who built a career at the University of British Columbia on the shoulders of his former Ph.D. advisor, Dennis Chitty, asked back in 1972: "How individualistic are populations? If we can understand spruce budworm outbreaks in New Brunswick, will we also understand them in Ontario?"[26] Doubt was surfacing even back then.

That doubt brings us, personally, full circle. As mentioned earlier, when I arrived at the University of British Columbia for Ph.D. research, I was perplexed by the various "camps" established around different population theories. So I went to see the dean, Ian McTaggart Cowan, a respected and acknowledged leader in wildlife ecology in North America, to ask him if I had any future at UBC, because I could not see myself fitting into any of the camps. He reassured me by saying that he, too, found himself fitting into no single camp. His thinking was that, just as individual species have evolved different mechanisms and behaviours to win at the evolutionary game, those differences must be reflected by a wide variety of ways that their populations are regulated.

At the time, this was good enough for me, and after our own journey of several decades in population ecology, it still is. The field has arrived at a relatively mature set of ideas. With no single mechanism to explain how all species are regulated, we have at hand some excellent frameworks to help determine what regulates whatever species in whatever situation we want to understand – give or take a dose of humility.

Standing back from the detail, it is apparent from this and the preceding four chapters that all populations are either regulated directly by the environment, or indirectly by internal mechanisms that the environment triggers. In either case, population control is a necessity for biological order on Earth.

FOOTNOTE

WHAT ABOUT HUMANS?

The single most important species whose populations we need to understand is our own. Among the many ways populations are regulated, where does the ultimate fate of humans lie? An ultimate fate there must be, because "no species can expand indefinitely in a finite space." For our species, will the predominant regulator be an extrinsic influence: food shortage, disease, air or water pollution? Or will intrinsic mechanisms, triggered by overcrowding, cut in earlier: physiological stress, wars, riots? Will our fate be an eruption and crash scenario, or a cycle? Have we passed the possibility of stabilizing at a long-term, sustainable size?

If you are an ardent optimist, then you might want to skip ahead and avoid this section. However, if you want a realistic ecological perspective on the application of population principles to humans, then read on.

The global growth curve of humans deviates from the normal S-shaped one, due to a protracted early slow rate of increase. That early section of the curve appears almost flat because the vertical axis must be tall enough to accommodate the current world population of more than 6 billion people. Nonetheless, the curve had risen sufficiently by 1798 to alarm Thomas Malthus, who saw it approaching the carrying capacity at that time. After that date, however, the agricultural and industrial revolutions caused an unprecedented increase in carrying capacity, and we experienced an explosive, essentially eruptive increase that reached the point of inflection on the graph and began very gently to arch over only in 1987. That is when the annual increase in population size at last began to decline (the population size continued to increase, but by fewer people each year).[27]

This event is one of the most significant in our history, because it was inevitable, as it is in all populations. It happened because the annual rate of increase, analogous to an interest rate, began to decline fifteen years earlier. Thus, the year 1963 is the one we should celebrate, because after that, a decline in the annual increase in population size was inevitable.

The annual increase in population size did not decline immediately. By analogy, an interest rate may drop, but the annual size of a financial portfolio may continue to increase for a while, because each year the size of the portfolio to which the interest rate is being applied has grown, albeit in declining amounts.

The drop in the population "interest rate" since 1963 has been dramatic, from 2.2 to 1.15 percent, but, as explained, the drop in the annual increase in population size has been less so, from 86 million in 1987 to 77 million in 2008. Now, with 4.2 births per second and 1.7 deaths, the net rate of increase is a "modest" 2.5 people per second, or approximately 220,000 per day.

This decrease in the world population growth rate beginning in 1963 has been due exclusively to a decrease in birth rate that began at that time, especially in developing countries, and has accelerated since then. In the developed countries, birth rate began to decline earlier, before 1900. However, offsetting the effect of declining birth rate on slowing population growth has been a decline in death rate (increased longevity), most dramatically in developing countries beginning just after World War II. This decline in the death rate reached its greatest extent about 1970. Since then, in both developing and developed countries the death rate has fallen only slowly.

Most remarkable about this initial decline phase of the human population has been its difference from the analogous phase in wildlife species, where birth rates tend to be maintained near a biotic potential (or the maximum possible), and where population decline is driven by an increase in death rate due to density-dependent environmental factors (those that exert greater force with greater population density). Instead, our initial decline phase is a result of our unique biological trait of foresight. The individual option to choose the number of offspring, facilitated by advances in birth control technology, increased education and wealth, and a desire to "live a good life" has caused the global drop in birth rate. And global death rate, instead of increasing with population size, has been decreasing due primarily to medical and agricultural advances.

These are global statistics; in certain places, notably Africa, the rates are considerably different and much closer to the norm for wildlife populations. There, birth rates are still high, and environmental resistance is increasing the death rate. Several extrinsic factors are involved, all of them density dependent, and all compensatory (if one thing doesn't cause death, something else will): disease, especially AIDS; air and water pollution; sanitary conditions; and the availability of food.

Current projections are for a peak human population at between 9 and 12 billion, arriving sometime in the latter part of this century. This estimate is based on a simple projection of the rate of change in birth rate and the extant or diminishing death rates since 1963. The calculation is simple in concept but complex in reality because of wide regional differences. And it has one important flaw. Because environmental resistance working to increase death rate has not yet happened globally, its inevitable effect could not be built into these forecasts. For any species, incorporating environmental resistance is essential to construction of a population model. But we have insufficient evidence to draw the shape of the upper part of the human population curve.

It is unrealistic to think that we could simply arrive at a carrying capacity without the contribution of any increase in death rate. The reason is that increasing population size itself increases its probability. By very definition, density-dependent mortality is turned on, or invoked, by population density. Besides, in many parts of the developing world, which includes a large part of the world's population, the direct evidence is already there. Human populations there are paying the price of increased mortality.

It is also unrealistic to believe that the developed countries will not have to pay, too. In all populations, density-dependent mortality is a logical necessity, as Lack and Nicholson postulated decades ago. We have such a rich menu of possible causes: extrinsic factors operating as mega-disasters – flooded cities, global food shortages, epidemic diseases, nuclear war; intrinsic factors acting subtly – stress-related diseases, crime, social breakdown, economic meltdown. Already in places such as Russia longevity has decreased, due primarily to high adult mortality in older males from cardiovascular disease and in younger males from accidents, suicides, and murders.

Moving towards global population stability would require continuing to make the birth rate decline, even considering the optimistic scenario of the death rate holding its own for a while at its current level. If instead we were to just stabilize the birth rate in the next few years, rather than decrease it, an annual increase in population size would still occur, because, as mentioned, a constant growth rate applied to a population still allows that population to grow.

If the birth rate in developing countries declined to its current level in developed countries, the world population would stabilize, especially with a little help from an increased death rate resulting from an older age structure. Then, at whatever world population density that occurred, the average quality of life index, or what many countries of the world might more realistically call their average misery index, might be sustained, even without the major increase in death rate brought on by density-dependent environmental factors. "Sustainable society," a term bandied about these days as a utopian objective, cannot mean any better than that, unless we achieve a reduced human population. Maybe we in privileged parts of the world think we can sustain what we have and let others bear the consequences. But ultimately, with regional immigration and emigration, we humans on planet Earth are one population.

Most population ecologists would agree with E.O. Wilson, who wrote, "It should be obvious to anyone not in a euphoric delirium that whatever humanity does or does not do, Earth's capacity to support our species is approaching the limit."[28] It seems unlikely that the biosphere can maintain the current 6.7 billion people indefinitely: provide the ecosystem service; absorb the toxins; counter our climatic effects; supply the resources; let us win the evolutionary arms race with epidemic diseases.[29] And if we include quality of life in the concept of carrying capacity, which seems an obvious necessity, and consider the environmental impact of extending its current level in developed countries to developing countries, then it is even more probable that we are in the situation of chronic population overshoot already.

However, we have an option not available to any other species: to exercise foresight – which, under the circumstances, is our most valuable biological attribute. The fact that the reduced global population growth rate experienced since 1987 is almost entirely due to employing foresight, which has reduced the birth rate, albeit for personal rather than global population reasons, shows that, *biologically*, it is possible.

PART 3

ECOSYSTEMS: WILDLY ACTIVE

THE DIFFERENCE BETWEEN a landscape painting and a landscape is that the former is static while the latter is wildly active – living things transforming solar to chemical to mechanical energy and handing that energy off from one trophic level to the next – genes being selected, activated, discarded – plant roots, fungi, and bacteria probing the soil for nutrients – species laying down ecological pathways and connections, adapting to one another, forging multitudes of relationships – populations ever adjusting to environmental contingencies – birth, death, decay, rebirth . . .

All this action sounds like a recipe for chaos, like fifty kids running wild in a daycare centre with no supervisor. But ecosystems are highly ordered, despite being on top of the biological pyramid and incorporating all the complexity of all the lower levels, right down to the gene.

How do ecosystems do it, and is a supervisor even present?

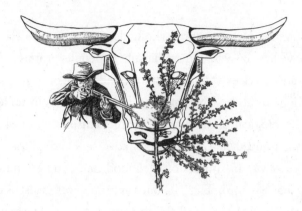

CHAPTER TEN:

WAR ON THE SHRUB-STEPPE

Across vast sections of the west, from the southern interior of British Columbia to central California and Nevada, where the scenery is reminiscent of old cowboy-and-Indian movies, guns are (metaphorically) still blazing. Here is a raw and ragged land decked out with coulees and canyons, mountain peaks and plains, and beauty worth fighting for. It is a land inhabited by ecological desperadoes. Life and death turn on a slim, competitive whim.

THE BRUTAL, ONGOING WAR on the shrub-steppe is an ecological war. The battles fought here are between ecosystems, entire assemblages of species marching together to the drumbeat of ecological connections. When one species is killed, especially a dominant one, then death may cascade through the ranks. Too often, these death cascades have been triggered by aliens – invaders from a foreign land.

Most people have failed to notice this war because we are not ecosphere people any more, wholly dependent on the ecosystem where we live. Instead, we are biosphere people, propped up with goods and services shipped from around the world. Just look in your fridge or at the label on your Blackberry or your underwear.

This ecological war is subtle – another reason that most people have not noticed it. Defraud your adversary rather than attack him; take away his means of livelihood; out-compete him. That is safer than open combat. The outcome, however, is just as deadly.

Like many human conflicts, this one has been over land and resources. Here, the coveted resource has been water. In the past, water rights have strewn destitute settlers and ranchers around the west; more recently, it has extirpated species.

Most western ranchers have known about this war, even agonized over it, whether realizing or not their own complicity. They certainly have felt the effects – right in their pocketbooks. Cattle, you see, do not do well on the land they degrade.

We knew nothing of this war until we moved west in 2000. Then, unexpectedly, we were thrust into it because of what we initially thought was a simple process of stimulating the creation of a national park in the Okanagan and Similkameen valleys of south-central British Columbia. This region is a northern extension of the cowboy country that lies mostly south of the B.C.–Washington State border. Suddenly we discovered the war at our doorstep.

Parks Canada had no national park in what it calls "Natural Region 3 – the dry interior of British Columbia." However, momentum built until, in 2003, the federal and provincial governments signed an agreement to evaluate a park's feasibility. A representative from Ottawa arrived, armed with little ecological understanding of the dry interior. Perhaps to promote peace, he told people that virtually any land use except logging, mining, and new towns might be acceptable in a new national park. At a public meeting the ranchers asked, "What about cattle?" His answer was an ambivalent, "Maybe."

Oh, really? In a national park the integrity of the ecosystem is supposed to come first. How could the option be left open to give cows room to do their metabolic and digestive thing there, too? Were parks officials singing the old refrain?

> *Oh ranchers, oh cowboys, join hands in a ring,*
> *Dance on the tailgate, spy birds on the wing,*
> *Give vent to aesthetics and join as we sing,*
> *A cow in a park is a beautiful thing!*

The southern interior of B.C. is not good ranching country – too dry. Farther north is better – cooler temperatures, higher dew point, more moisture. Farther south is largely better too, for the same reasons, but caused by higher altitude rather than latitude. Here, however, it takes about six hectares (fifteen acres) to support one cow. The climate is too dry due to the intense rain shadow caused by the Cascade Mountains immediately to the west.

Yet environmental conditions never seem to have curtailed cattle introductions. Except in the far north, wherever there are a few strands of grass, invariably there are cattle. We humans, if viewed by aliens from outer space, might be considered a "cattle culture." It might be difficult to determine whether humans husband cattle or the reverse. Cattle make up the greatest biomass (the number of animals times the weight per animal) of any mammal on Earth and exceed human biomass by about one-third. Worldwide, we tend almost 1.4 billion cattle.[1] Permanent pasture is the most widespread land use in the world, covering 3.4 billion hectares.

The central reason for making this new national park in the Okanagan region was to prevent the butchery of dry country by cattle. Because this ranching issue threatened to undermine the entire initiative, we consulted our shelf of ecology books and their vegetation maps of the "intermontane." The intermontane is big – all the land between the Rockies on the east and the Cascades and Sierra Nevadas on the west, extending from southern British Columbia, where it fingers into the Okanagan and Similkameen valleys, to southern California. Most of this big-sky country is good scenery for a John Wayne movie. In Washington State it includes the vast Columbia Plateau, a lava outpouring 15 to 20 million years ago that raised the landscape to about 1,000 metres and flattened its relief.

On the maps we found a curious inconsistency. Some showed the Columbia Plateau and Okanagan as "shrubland" or "cold desert."[2] Others mapped it as "bunchgrass" or, alternatively, as "Palouse prairie."[3]

Ignoring the reference books, people where we live like to call the Okanagan Valley "desert." The word has a dramatic ambience about it. Playing up the desert moniker even further, some people, particularly subdividers or golf course developers, use the more evocative but incorrect term "Sonoran." You can own a home in stunning "Sonoran Estates," or tee off on the greens of sunny "Sonoran Links," or play bingo at the "Sonoran Community Centre." We even have a "Sonora Spa," where your desert experience comes from the radiance of a heat lamp.

Technically, calling the region's lowlands "desert" is acceptable – "cold desert," that is, because it is dominated in most places today by shrubland, and grasses are only subdominant. On grasslands or prairies, the reverse is the case. Farther south, in southern California, are the hot deserts, Sonoran, Mohave, Chihuahuan, that are characterized by large succulent cacti. The only cacti in the Okanagan are the eight-centimetre-high (three-inch) prickly pears that grow across Canada. Large cacti are found here only as frescos adorning the walls of otherwise ugly buildings.

To determine which designation in the ecology books was correct for the Columbia Plateau to the south of us – shrubland or grassland – we loaded our pickup and set out. We did not realize that we were entering such a war zone.

Being at a higher elevation, the Columbia Plateau is cool, which should have made it good for grass. Instead, to our surprise, on the level ground we found wheat, in places stretching as far as the eye could see, no different from southern Saskatchewan. Whatever native vegetation had grown there was long gone. Occasionally, this "wheatland highway" dipped off the flat lands and crossed wide, rock-strewn valleys left behind by some of the largest floods the world has ever known. At the end of the last ice age, a series of deluges took place that carved these wide valleys out of the foundation lava rock. These are the scablands of Washington State, eroded by thirty or so gargantuan torrents. When the ice dams repeatedly broke in northern Idaho, they suddenly released volumes of water as great as Lake Ontario.[4] They scoured the land in swaths trending southwest across the state on their way to the Pacific Ocean. What grows in these places today? Predominantly big sagebrush, *Artemesia tridentata*, the ecological victor. Very little grass.

We continued south through the Blue and Strawberry mountains that run east-west in northern Oregon, and into the high desert region of southern Oregon. What grows there? More sagebrush. At Mono Lake in northern California we bought a book called *The Sagebrush Ocean* – an apt title to describe the ecology of this huge, dry land.[5]

What, then, of the "bunchgrass grasslands" or "Palouse prairie" marked on the maps? Where had it gone? On a map of Washington State, we spotted the intriguing designation "Arid Lands Ecology Reserve," situated alongside the meandering Columbia River south of the Grand Coulee Dam. There, indeed, we found rolling hills of grass. Elated, we hopped a fence to take some pictures, ignoring the signs saying "Danger, Keep Out." Later, in doing some web-based research, we learned that a series of three consecutive fires had swept the area. Repeated fires can remove shrubs, especially if they occur before the replacement shrubs are old enough to produce seed.

In our web search, we also learned why the signs around the area had, in small print at the bottom, "U.S. Department of Energy." Maybe we should have heeded them, because this area is part of the Hanford Nuclear Site, one of the largest nuclear waste dumps in the United States. This site has provided a "safe house" for the most high-level nuclear waste in the western hemisphere, much of it still buried in underground tanks that leaked for decades.[6] After seven years of the most intensive cleanup in history, costing more than $2 billion per year, the U.S. government admits that 25 tons of plutonium "still await disposal." They are kept under constant armed guard.[7] In a masterful effort to put a good spin on all this, a web page of an "Ecology Group" funded by the Department of Energy states, "Scientists representing various disciplines can . . . resolve questions concerning the responses of this large, unique and fragile landscape to the use of toxic pollutants, and other types of perturbations associated with humans."[8]

There were other candidate grasslands to check out. Near the Washington State – Idaho border is the town of Palouse, along with the Palouse River, and Palouse Falls – a logical place to look for Palouse prairie. And there it was, a magnificent grassland dominated by Idaho fescue and bluebunch wheatgrass stretching back expansively from the canyon rims. We camped in a small state park by the falls and in the early evening took pictures of the intense sunlight flooding across the colourful basaltic rocks and golden grasses.

We found more grassland, too, in the gorges of the nearby Snake River, into which the Palouse River flows. There, the grasslands are confined to the sides of valleys too steep for combine or plough. Where the land was flatter, however, it had been converted to wheat.

The last grassland landscape we found – big, impressive, rolling hills of it – was in northern Oregon on the southern edge of the Columbia Plateau near the John Day system of rivers. But this area, together with the Palouse, Snake gorges, and Hanford, made up all the extensive grasslands we were able to discover. Elsewhere, the Palouse prairie has largely disappeared. Textbooks that still call this large, northern region of the intermontane, including the Okanagan and Similkameen valleys, by that name – and some textbooks do – are out of date.

What happened? Had shrubland simply out-competed grassland, and if so, why? Competition between species presents an ecological conundrum. It has been studied in detail because of its huge implications for both evolution and ecology. Darwin emphasized competition both within and between species as a driving force in evolution. He observed, "Each species, even where it most abounds, is constantly suffering enormous destruction at some period of its life, from enemies or from competitors for the same place and food; and if these enemies or competitors be in the least favoured by any slight change of climate, they will increase in numbers; and as each area is already fully stocked with inhabitants, the other species must decrease."[9]

Ecologists today agree about the importance of competition between species, describing it as a fundamental force. It determines the role each species plays in its community and causes the partitioning of resources that underpins biodiversity. Some ecologists have gone even further and enshrined interspecific (between species) competition with the undeniable authenticity of mathematics in the "Lotka-Voltaire equations." These two equations, one for each of the competing species, assert that the change in population numbers from year to year is a consequence of each one's birth rate and death rate multiplied by the negative effect of the adversary species. Both species are losers because the other uses resources they need in common.

Yet, remarkably, scientists have had difficulty in actually demonstrating interspecific competition, either in controlled laboratory experiments or in nature, in plants or in animals. This difficulty does not mean that interspecific competition has been overrated. Instead, the scientific method may come up short. Consequently, interspecific competition has been called "community ecology's most celebrated impasse."[10]

Several arguments, however, support its importance despite the lack of experimental evidence. Most important is the "ghost of competition past" theory, which postulates that competing species found ways in the past, soon after they first encountered each other, to minimize direct competition.[11] Competition is risky. Much more is at stake than a ribbon or a medal. Accordingly, competition is reduced by a "niche shift," a tricky little evolutionary manoeuvre.

Take, for example, two nondescript species, sage sparrow and

Brewer's sparrow, that supposedly compete with each other. Both like the same dry, open sagebrush country, eating seeds and insects gleaned from foliage and bark. Potentially they compete for both food and habitat, each one's population suffering because of the other. But natural selection in such a situation will favour any adjustments in behaviour, habitat requirements, or diet that will minimize competition. Adjustments take place automatically, and invariably, through selection from the range of individual traits in the population. While sage sparrows and Brewer's sparrows overlap considerably in their choice of food, competition has driven a niche shift so that the former forages more consistently from the ground, the latter from low shrubbery.[12] For many species, this competition-driven ecological compromise makes coexistence possible.

This explanation makes intuitive sense in the way it accounts for the many fine differences in niche shown by species living together in an ecosystem. But because it is based on past events, only rarely can it be "proven" by observation. Here is a classic situation in which the requirement of science to be based on experimental evidence proves inadequate. Experimental evidence is obtained more easily when studying lower levels of biological organization – genes, tissues, organs, organisms. But for higher levels – species, populations, ecosystems, landscapes – it is impossible to isolate any one feature to determine its significance. Instead, a search for pattern or an observation-based, logical explanation that fits the evidence are both accepted, although less powerful, alternative scientific methods.[13]

Yet, in other ways, interspecific competition may be working but undetectably. It may operate only occasionally, such as in years when the weather causes food shortages, or cease when

populations are temporarily depressed by influences such as predation, disease, social stress, or weather to levels where they no
longer compete.

Bearing in mind these caveats on competition, then, are shrub-
lands and grasslands truly competing across the vast intermontane
west? Or have they reached some permanent or periodic standoff? Looking to the plants themselves for answers, we find that in
the struggle for moisture, sagebrush has some winning features: it
comes armed with deep roots, and wizardry like shedding its
leaves in dry weather, and a high concentration of salts in cell sap,
which creates a negative pressure and helps it retain water. As
well, sagebrush leaves are densely hairy, reducing air movement,
which limits transpiration. Big sagebrush may practise chemical
poisoning, too, killing competing species in its immediate vicinity.
This weapon has been demonstrated under controlled laboratory
conditions but not yet in the wild.[14] All these adaptations may
reflect a niche shift made in competition with grasses.

Grasses, in comparison, developed in the intermontane almost
exclusively as "bunchgrasses," featuring distinct moisture-holding
clumps. Their clumped roots probe the soil for water but not as
deeply as sagebrush. As a result, sagebrush, and other shrubs such
as antelope brush and saltbrush, normally out-compete grasses
where conditions are dry.

These adaptive differences between sagebrush and grasses
developed long ago. Does that mean that, instead of feuding, each
one now ignores the other and just does the best it can with the
adaptations it has? Has a truce been arranged; the war over?

The answer was both revealed and symbolized in our next
dragnet. We spotted the name Butte Valley National Grasslands
on a map, located in northeastern California. Although National

Grasslands are managed by the United States Forest Service, we expected to find an intact grassland. The place took a while to find, because the entire region was covered either with sagebrush or, at times, two more salt-tolerant shrubs named greasewood and saltbrush. Finally, on a back road, we located the sign, but unfortunately, we had come several decades too late. The landscape was no different from the surrounding country – covered in sagebrush. Standing in the sagebrush beside the sign were two rather guilty-looking cows, as if designated by their kind to offer us an explanation. Crowning this irony, the only grass visible was at the base of the signpost – a clump of cheatgrass, introduced scourge of the west, perhaps the most prolific, extensive, and destructive of all exotic species ever to invade North America, and spread primarily by cattle.

Cattle, however, have answered the question of whether competition between grasses and sagebrush is still ongoing. The answer becomes apparent only with some historical background.

Historically, permanent grazers were absent in the intermontane. When one thinks of grasslands and prairies, one invariably thinks of bison (colloquially but incorrectly called buffalo) – thousands of them spread across a sea of grass, or a huge pile of their skulls dwarfing a man standing at its base. But bison did not inhabit the intermontane in any significant numbers over the past several hundred or perhaps thousand years. The evidence is in the journals of the early explorers. While crossing the Columbia Plateau, they almost starved. In 1804, Lewis and Clark, for example, had to eat their dogs.[15] In 1824, Hudson Bay traders "were reduced to eating their horses and stealing beaver meat from each other."[16] Chief fur trader Peter Skeene Ogden, who repeatedly crossed the Columbia Plateau, wrote in 1826, "We have now traveled upwards of 200 miles and not a track of animals to be seen, we have endured privations greater than I am willing to relate."[17] John Fremont, who laid out the Oregon Trail for immigrants in 1841, wrote, "Once again after all the months and the great distances, there were buffalo in sight."[18]

Archaeologists have searched the Columbia Plateau for bison bones. On the east side, close to the Rockies and Bitterroots, evidence reported in 2003 suggests that "small herds were relatively common between 2,500 and 500 years ago."[19] There, bison could have moved into the mountains during the hot, dry summers. Farther out on the central and western portions of the Plateau, no bones have been found. Draw a line along the Idaho – Washington State boundary and you have a good approximation of the western limit of plains bison.

Not that they were blocked by mountains. They could have made it through several low passes in the Rockies, especially by following the Snake River Plain just north of Salt Lake City. Records of trappers show that they actually did use this route.[20] One sighting even put a few bison out on the dry plains of central Oregon near present-day Harney Lake, which is the westernmost occurrence in the United States.[21] But they did not persist.

Why bison failed to exploit this abundance of intermontane grass has been the subject of much speculation. One theory holds that grasses growing there were, and are, protein-deficient for the bison's metabolism, due to lack of summer growth.[22] During growth, most plants convert more of the original product of photosynthesis, glucose, into protein to make the protoplasm of new cells.[23] Growth of most intermontane grasses, however, ceases during the summer due to insufficient moisture.

An alternative explanation for the lack of bison is that the deeper snowfalls in the western part of the Plateau were a hindrance. Bison graze by moving snow with their heads but can do so only up to about seventy-five centimetres deep.[24]

The most controversial explanation is found in a scientific paper addressing the question of why Lewis and Clark almost starved when they crossed the Columbia Plateau. The authors proposed that wildlife overkill by humans was to blame. The population of native people was unusually dense, propped up by abundant salmon and other fish in the rich waters of the Columbia River and its tributaries.[25] The by-product of the high human density may have been overharvest of bison.

The authors of this overharvest hypothesis went on to suggest that the grasses were of sufficient quality on the Columbia Plateau to support both bison and elk, and as evidence cited a population of introduced elk that thrived for more than twenty years in the Hanford site. However, on further investigation, we learned that these elk took advantage of nearby cropland to forage, and so do not provide legitimate evidence for the adequate quality of the native rangeland. Now the elk are scattered, if there at all.

More likely than the overkill hypothesis is a train of events: low summer precipitation results in low growth rate of grasses, as well as associated low levels of protein (and often-related low palatability), which both kept bison from living in the intermontane and prevented the grasses from being subjected to heavy grazing that they could not withstand. And with this negative feedback to keep grazers out, the system went happily on – until cattle arrived.

In contrast to the intermontane, on the Canadian and American central prairies, bison thrived. So did black-tailed prairie dogs that once numbered in the billions.[26] Prairie dogs, too, have never lived in the intermontane. Why were all these grass-eaters supported by prairie grasslands but not by intermontane grasslands? The reason is simple: not all grasses are created equal. The prairies grow a preponderance of rhizome grasses rather than bunch-grasses. Rhizome grasses are what you plant for a lawn. They have shallow roots called rhizomes that, being shallow, are ready to latch onto any sudden precipitation. Summer thunderstorms on the prairies are common. Rhizome grasses respond quickly, green up again, fix more energy, and so provide more fodder for grazers. However, because of summer drought in the intermontane, rhizome grasses are uncommon, as their shallow roots are unable to suck as much deep moisture from the soil.

Bunchgrasses respond to summer rains too, although somewhat more slowly. The three thousand or so bison in Yellowstone National Park today do fine on primarily bunchgrasses, but Yellowstone's bison range is mainly in high country, two thousand metres or more above sea level, and it normally receives summer rains. Bunchgrasses in the intermontane, however, if heavily grazed repeatedly in spring, which is their critical time for energy capture, succumb. Moreover, lacking rhizomes, they cannot re-sprout, so they are unable to recover from grazing that removes their seed heads.[27]

And so, in the mid 1800s, cattle were brought to the grasslands of the Columbia Plateau and the Okanagan. While our question – are grasslands and sagebrush still in active competition? – was not likely foremost in the minds of the ranchers who brought the cattle, they did run an elegant experiment. The lands had not supported a permanent low-elevation grazer and were not adapted to heavy grazing pressure. The only large native grazers – mountain sheep, mule deer, and possibly caribou – were altitudinal migrants, moving up into the mountains in summer instead of eating in the same lowland places all year.

The experimental results? Repeated grazing by these bovine immigrants put grasses at a competitive disadvantage with their ancient enemy, big sagebrush. On former grass-dominated landscapes, sagebrush capitalized on the weakening of its competitor, and the ratio tipped dramatically towards sagebrush, today the most abundant shrub in North America.[28] Here is a clear demonstration of interspecific competition happening now, one of the most dramatic examples known.

To be fair, some range ecologists cite contrary evidence from experiments in which cattle were fenced out of small areas and sagebrush did better inside than outside the fence. Ecosystems, however, are influenced by many environmental conditions operating in interactive and simultaneous ways. On small parcels of land, even well-designed studies can lead to results that may not be widely applicable. Operating everywhere are differences in microclimate, soils, slope, aspect, insects, and other invertebrates. Only a broad, landscape-wide perspective shows the true picture. That picture is an overwhelming increase in sagebrush across intermontane rangelands.

There is another influence on the dominant vegetation; nature rarely deals in simple relationships. "Screw-up variables" they are often called by frustrated field ecologists. The exclusion of fire can favour shrubs, too. Fire is a natural event on grasslands, and it usually helps perpetuate grass. It does so because big sagebrush, after fire, seldom re-sprouts from its roots but must depend on slower-acting reseeding.[29] If fire sweeps a landscape twice – the first time removing the mature sagebrush, the second time removing the seedlings – then the remaining seed bank in the soil may be insufficient for another sagebrush crop. How much of the conversion of grassland to sagebrush in the northern parts of the intermontane is due to control of wildfire rather than cattle is difficult to determine.

Cattle are far from exonerated, however, for messing up the intermontane. Nobody contests another tragic influence of these "hoofed locusts." They have been, and still are, a major distributor of the most destructive exotic plant ever to invade North America: cheatgrass. Cheatgrass is native to the Asian steppes and Europe. It was introduced repeatedly to North America, first about 1900,[30] then at least six more times, according to genetic studies.[31] In its native land it evolved with, and became resistant to, heavy grazing from wild horses and other species.[32] In the intermontane of North America, cheatgrass has simply out-competed native bunchgrasses that were subjected for the first time to intense cattle grazing only in the late 1800s. Like other invaders, it brought its existing adaptations with it, ready-made to out-compete the residents. It is an annual grass, germinating quickly in response to soil moisture that usually comes in spring and fall, allowing it to produce at least two seed crops each year. It can overwinter as a seedling, giving it a jump-start in the spring; its roots grow rapidly, commandeering moisture from more slowly growing bunchgrass roots and allowing earlier maturation;[33] its seeds can persist in the soil for many years.[34]

Interspecific competition acts in the present in two situations: when an alien species invades an area[35] or when native species expand or shift their ranges. In either case, there is no ghost of competition past because there has been no past association. Cheatgrass, as an alien, therefore, is in active combat with bunchgrass today. It has been, and is, helped by its accomplice, cattle, and to a lesser extent horses and domestic sheep. Their hooves make disturbed ground available for colonization, more than

400,000 square kilometres (1,544 square miles) of it in the intermontane.[36] In nearly every place where cheatgrass thrives, its
establishment is associated with disturbed soil.[37] Moreover, livestock grazing has weakened the moisture-stressed native bunchgrasses. And the process by which cheatgrass spreads is simple:
cattle eat the seeds, walk somewhere else, and defecate. Magnify
that by many cattle, a vast extent of land devoted to them, and a
century of time . . . here are all the makings of an ecological
catastrophe.

For ranchers, ironically the unknowing sponsors of this plant
immigrant, its spread has been disastrous. Cheatgrass is not a preferred forage species for cattle. Except during green-up, its nutritional value is very low.[38] Cheatgrass burns more readily than
native grasses, lessening the periodicity of fire to its own benefit.[39]
Spring fires, especially, favour cheatgrass regrowth over native
bunchgrasses, magnifying its dominance.

In short, cheatgrass has proven to be an ecological curse, a ter-
rorist on the shrub-steppe, a ruthless plunderer. It has claimed so
much territory that it can never be defeated. In the Okanagan, the
best we can do now is to make some of the land into a national
park, remove the livestock, and let the healing processes of nature
achieve whatever they can. Without the bare ground being
churned up by cattle hooves, native species may be able to hold
cheatgrass in a competitive standoff.

Cattle and cheatgrass demonstrate that the self-organization
of an ecosystem can be grievously wounded. The incapacity of
one element may lead to the failure of others. For instance, with
the decline in native bunchgrass habitat, birds like grasshopper
sparrows and burrowing owls have declined too, and both are now
listed as "of special concern" or "at risk" under endangered species

legislation in the United States and Canada respectively. After all, ecosystems are *systems*; their components interact. Because of cattle and cheatgrass, the naturally ordered ecosystem integrity of the intermontane has taken a hit. While the system still functions in an impaired way, the damage disturbs any person who can recognize what has happened. For those who do not even know what cheatgrass is, and would not care or might turn a blind eye for economic or personal gain, there is no environmental cost to bear. As written by ecologist Aldo Leopold, "One of the penalties of an ecological education is that one lives alone in a world of wounds." Much of the damage inflicted on land is quite invisible to laymen.[40]

And what of the national park and the political struggle on the shrub-steppe? This is no inconsequential skirmish. The objective of national parks is lofty: to foster native species and ecosystems, and allow the forces of nature to dominate as much as possible in today's world. These are the very forces of life support that provision the biosphere, provide ecological services, and stabilize its cycles. Because we understand the foundations of nature only dimly – witness the "celebrated impasse" over interspecific competition – isn't it prudent to give over a significant portion of the planet to their free play? After all, they have served life quite well over a few billion years.

The national park movement is worldwide, existing even in countries whose economic well-being is a pitiful shadow of North America's. Between 1980 and 2000, the amount of land in national parks or equivalent reserves in the world doubled to almost 19 million square kilometres (7.33 million square miles) – almost twice the size of Canada.[41] If our life support is to extend very far into the future, it will be due, in part, to this ideal of setting aside land where humans allow nature to play its obviously successful role.

None of the shrub-steppe ecosystem in North America is protected, except for a fragment in Nevada's Great Basin National Park. Instead, it has been incrementally carved up, bulldozed, cemented over, subdivided, and grazed by livestock. There will be no next chance. While cattle are one thing, condos are another. Ranchland may be partially restored, but not land that has been reconfigured for urban development. Housing is advancing everywhere on the open shrub-steppe landscapes – here a scattering of "ranchettes," complete with llamas and horses, there a cluster-development – all with a great view, all attractive to urban refugees. The character of both the Canadian and American west is changing fast. Wild West backdrops must now be filmed with care to avoid showing the trappings of civilization. At any moment an ATV may come over the hill.

The drive to establish a national park in British Columbia's Okanagan was given momentum by a species-specific conservation rationale: to save one thousand square kilometres of one of Canada's top four hotspots of species diversity, including the greatest number of endangered species. When this became known, public enthusiasm for the proposed national park grew, with more than twenty thousand people signing a petition. The debate should have been brief; success seemed certain.

Yet, six years after the feasibility study began, five federal ministers of the Environment (the ministry that houses Parks Canada) have come and gone, all seemingly as inconsequential to the process of establishing the national park as telephone poles slipping past a car window. The Parks bureaucracy is mired in the complexity of sorting out federal, provincial, First Nations, and local interests, and in working through a tangle of rights: land ownership rights, grazing rights, hunting rights, ATV and snow-mobile rights, horse riding rights, mining rights, water use rights – every conceivable kind of right except those of the ecosystem itself, which of course doesn't have rights at all, only consequences.

Finally, at an impasse, Parks Canada folded its tent and silently slipped away. It re-posted its project manager to Alberta. Officially, when queried, the feasibility study is "on hold," a disturbingly ambiguous classification, but accompanied by a promise from Parks Canada that the file is still active.

Meanwhile, the alluvial fans are being converted to vineyards, "For Sale" signs appear along the roads, and open-land housing becomes established, forcing back nature like bulwarks on a bat-tlefield. The areas of remaining shrub-steppe await their inevitable, human-reconfigured fate.

CHAPTER ELEVEN

BEHIND THE SCENES

The teacher unexpectedly enters the classroom and is met with sudden silence. A knot of children hunch in one corner. Several more stand on their desks. Others are up at the blackboard, chalk in hand.

"What is going on here?" she demands. Silence. None of the children will own up. To find out, she will have to ask probing questions, consider the classroom dynamics – the leaders, followers, agitators – even possibly isolate some of her subjects.

A forest, a mountainside, a meadow are like that classroom, and we, like the teacher, ask: "What is going on?" The trees and wildlife are silent. It takes patient sleuthing to find out.

THINKING ABOUT PROCESSES can be achieved only by habit, because species are so obvious and captivating themselves. But slipping down the rabbit hole and squeezing through the door into this other realm, like Alice entering Wonderland, is worth it. It adds a dimension to ecological understanding.

Living things are always doing something, and thereby unwittingly playing out processes. Plants are photosynthesizing, transpiring, extracting soil nutrients, metabolizing, growing, budding, flowering, distributing pollen, fruiting, reproducing. Animals are foraging, preying or being preyed upon, competing, mating, rearing young, receiving or giving care, defending land, migrating, moulting, metamorphosing. Even the inanimate parts of ecosystems are active: elements sifting through sedimentary and gaseous cycles; water moving through hydrologic regimes; soils of shorelines eroding and being re-deposited; weather systems playing over the land. Molecules, in repeated resurrection, shift from non-living to living and back again. Species live out their life cycles, then pass the torch to their offspring. All this activity is happening in any ecosystem in a marvellously coordinated way.

Seasons help do the coordinating with variations in temperature, moisture, and/or light. The importance of seasons is why relatively sudden global climate change has such potential to disrupt the finely tuned adaptations of life. Helping coordinate ecosystem activity, too, are day-night cycles, which have induced circadian (twenty-four-hour) rhythms in the physiology and behaviour of many species. Also contributing to ecosystem coordination is succession, whereby plants and animals cause physical changes at a micro level over short periods of time (see Chapter 13). But always providing the guiding hand is natural selection, weeding out the conflicting situations and reconfiguring the processes that do not work, until the whole assemblage of living things in any one place reaches an internal accommodation.

The north shore of Lake Superior is wild and rocky. Ocean-sized waves slap and suck against granite headlands. Wind-driven fog lashes the dripping green bulwarks of spruce and fir. Sunny days round off the jagged edges of storm fronts only occasionally, and even less often the wind quiets down.

Around a rocky headland, an undersized motorboat beat into the wind, spray flying, hull pounding the waves. At the helm, a man with cap backwards, jacket flapping, was intent on keeping the boat from slipping broadside in the swells. Lake Superior is unforgiving to small boats. Soon he made it to the slack water of the bay and cruised up to the sandy shore, where we waited.

Bob Severs, the driver of the boat, was a long-haired, lanky, jeans-clad student, a product of the post-hippie era, who was about to prove himself by tackling a particularly vexing problem that Parks Canada had put out to tender. Parks Canada's problem was caused by one of its own desk-bound planners. As part of a master plan for their new Pukaskwa National Park, the planner had proposed a major campground in a controversial, or maybe more accurately ridiculous, place known as Oiseau Bay. To reach the bay would require a road cut thirty kilometres through superb wilderness that hosted its own isolated herd of woodland caribou and a dense moose population.

The bay was beautiful, with a long, crescent-shaped sandy beach, rare on the rock-strewn north coast of Superior. A sparse mantle of beach pea and horizontal juniper stabilized the back dunes, plants that could easily be pummelled by sports such as frisbee throwing and volleyball. Clinging to the sand in one proposed high-traffic area was a rare colony of pitcher's thistle, a species confined to a few suitable shoreline sites on Lakes Superior and Huron. Adapted to sand dunes with long roots, it produces seeds only once during its decade-long lifetime, then dies. Behind the beach lay an extensive jack pine flat, and off to the northeast a black spruce bog, both destined, according to the planner, to accommodate drive-in campsites.

A black spruce bog is not entirely – not even remotely – appropriate for a campground, a fact that even the planner knew, but he had never visited Oiseau Bay and was not particularly adept at reading aerial photos. Some of his colleagues thought that Oiseau Bay should be designated a "Special Area" and preserved for its remoteness and beauty. Therefore, Parks Canada had contracted out a two-year environmental impact study that became the focus of Bob's master's thesis. During his study, the value of seeing behind the scenes to what made the ecosystem work became apparent.

Bob spent the first winter taking courses at the University of Waterloo, then raided our storage locker for field equipment and headed for Pukaskwa. Parks Canada lent him an outboard motorboat, and he made his way to Oiseau Bay. We visited him there, flying in by float plane from the town of White River, forty kilometres away.

He had established a neat little campsite designed to levy minimum impact, back from the shore on the edge of a rocky headland. His trail to the beach and another to a pit toilet were marked out with string, and he meticulously kept track of the number of trips he made. The organic soil was only a few centimetres thick, and he wanted to record the amount of trampling it could withstand before it became pulverized, exposing the underlying sand.

Blackfly season was in full swing when we arrived. We chose that season, flies notwithstanding, because that is also the prime time to survey breeding birds. On compass bearings we laid out transect lines, and each morning we recorded all the singing birds along them. It was pleasant work. Under the jack pines, a deep mat of Shreiber's moss, a species similar to sphagnum, carpeted the forest floor. Sunlight filtered in weakly through lichen-draped branches.

But the forest puzzled us. Normally, small spruce trees grow up in the shade of jack pines. They grow progressively larger, eventually looming above the pines and taking over. Here, however, were no saplings of any kind.

To predict the impact of a campground, we needed to know: why the absence of saplings? What was going on to make this stand the way it was? What directions was nature's stage manger providing behind the scenes? What processes were at work? Instead of asking the normal naturalist question, "What species are here?" we asked, "What is going on here?" Species are only the result of underlying processes that caused them in the first place, and continue to sustain them.

For his environmental assessment, Bob developed a chart with human activities listed across the top and ecological processes down the left side. Human activities included roads, tent sites, campfire rings, toilets, parking places, trails, and the accompanying high-impact areas behind campsites where dogs poop, playing children beat down the vegetation, and adults rip apart the trees for firewood. The ecological processes he listed were the agents of change that are universal in all ecosystems, for example, soil-based ones like erosion, plant-based ones like photosynthesis and succession, animal-based ones like predation and parasitism.*[1] Along with all-pervasive natural selection, these ongoing processes

---

* A more complete list of processes includes the following. Abiotic processes: soil erosion, deposition, and development; water fluctuations and changes in quality; micro-climate influences. Plant-related processes: photosynthesis, nutrient uptake; succession; decomposition; evapotranspiration; herbivory; competition; disease; pollination. Animal-related processes: energy and nutrient transfer between trophic levels; habitat selection; competition; co-operation; predation; parasitism; disease; movements to and from critical habitats; population regulation; and mutualism of various kinds.

shape living things and their environments, the core of what really matters in nature.

At Oiseau Bay, succession in the jack pine stand was stuck. It was not proceeding to a normal spruce-dominated climax forest. The reason emerged when we compared it with other stands growing farther inland along the Trans-Canada Highway through northern Ontario. They all displayed an understorey of regenerating spruce. But none exhibited the exceptional, mid-calf-deep, continuous mat of moss that grew at Oiseau Bay. Stringing these two observations together, it became evident that moss was effectively preventing any seeds from reaching the mineral soil, a requirement for germination. The deep moss, in turn, resulted from ideal growing conditions adjacent to the Superior shore at one of its wettest, foggiest places.

Another feature of the jack pine stand at Oiseau Bay was that the trunks of all the trees were almost the same diameter, showing that they had started growing at about the same time. We confirmed that fact by drilling some with a tree core sampler and aging them by their growth rings. The only possible explanation was regeneration after fire. That fire had killed not only the previous forest but the moss, exposing for a short time the mineral soil. Very soon after that, the moss had re-carpeted the forest floor and terminated any possibility of further re-seeding. Fire is common in the boreal forest, happening on average every hundred years.[2] The future of this stand, we predicted, was to stay as jack pine until it burned again, then regenerate as jack pine until another fire . . . indefinitely. In that way, the stand was unique in the national park system and worthy of preservation. Opening of the forest canopy to build the campsite, however, would allow sunlight to reach the ground, which would increase evapotranspiration, dry the site out, and kill much of the moss. Spruce would invade, and the uniqueness of the stand would be lost.

When the study was complete, we recommended that Parks Canada abandon its plan for a campsite on this bay and instead build it at a more resilient site near the park entrance, which eventually was done. They followed our suggestion of zoning Oiseau Bay as a "Special Area," giving it the highest degree of protection. And as for the planner who had started all the controversy by incorrectly interpreting the air photos, he was promoted to a less consequential, more senior management position.

The story of our adventures at Oiseau Bay does not end there, however. Two footnotes must be added.

At the end of his field work, Bob boated back to civilization, represented marginally back then by the little gold mining and logging town of Marathon. He intended to get a room in the town's only motel and get cleaned up. The owner took one look at his grisly appearance and refused him accommodation. Bob was angry but laughed it off and camped outside town. Then, after graduating the following year with his master's degree, he decided to change fields and enter medicine. A few years later he emerged as a general practitioner and took the job as the only doctor in – where else? – Marathon. One day the motel owner showed up in his office. But, true professional, Bob did not diagnose him with some rare, incurable disease; he kept his identity to himself, and the motel owner did not make the connection between the old Bob and the new doctor.

The other footnote is that we were wrong in our prediction that the jack pine stand would perpetuate itself indefinitely. We had an opportunity to return, twenty-eight years later, with our daughter Jenny, who by then was the biologist managing science

and wildlife ecology for the park. We had flown with her by helicopter to the south end of the park to help assess the effects of
moose on forest regeneration following a fire. After camping there
for a week, we were boating back up the coast to Marathon. The
day was sunny, the water relatively flat. We rounded a headland,
and there was Oiseau Bay in all its splendour – still a wilderness
with no campground. We landed for a few hours to inspect the
jack pine stand.

We had trouble finding Bob's old campsite because, ironically,
the foreshore had been remodelled by a major storm. It had shifted
the location of the little stream's outlet by almost a kilometre and
had built and erased sand dunes. Most of the pitcher's thistles were
gone. It was the jack pine stand, however, that provided the real
irony. The storm had toppled about half of the trees, which now
lay scattered on the ground. In the sunny openings, growing on
the rotting logs or where roots had torn up the moss and exposed
the mineral soil, we found spruce saplings.

Unplanned environmental events, unpredicted catastrophes,
contingency, surprise – all are part of the way ecosystems work.
They are nature's clever ways of broadsiding you.

On a July day in 1977, a few years after our Pukaskwa study, we
were re-supplying our caribou research field camp in the Yukon
with a trip to town. A cool breeze was blowing down Whitehorse's
main street, the sky grey, suitably chilly for what was going on –
an environmental showdown.

Whitehorse was playing host to hearings for the proposed and
controversial Alaska Highway Gas Pipeline planned to run
through the Yukon Territory. If built, it would be the largest mega-
million-dollar project in Canada's history. Local people with

plenty of concerns were pitted against money men from "outside" with plenty of prepackaged answers, or, as the local newspaper put it, it was "men in boots against men in suits." We counted at least thirty-three biologists in town and remembered our first visit to Whitehorse, when the full contingent of "biologists" in the Yukon was only one person, an ex-RCMP officer, quick with a gun, and so eminently qualified for the responsibility of wildlife management.

Even more bureaucrats than biologists were in town for the hearings, with no intention of letting a bunch of caribou lovers or rare plant enthusiasts stop a development of this magnitude. The media were there, too, from all over North America. The pipeline was to service the "Lower 48."

Day after day in a high school gymnasium the hearings droned on. Various biologists testified on where caribou migrated, raptors nested, and moose wintered. They were cross-examined by other biologists hired by the pipeline proponents to assert that what they had just heard was irrelevant because of the advanced engineering capabilities of the proponent. The script could have been written in advance.

One day, down the main street of town strode Charlie Krebs, an ecologist from the University of British Columbia. He was in the Yukon studying population regulation in snowshoe hares, a keystone species in boreal and montane habitats. His study area was a broad spruce-willow swath near the Alaska Highway at the south end of Kluane Lake. As he approached we could see that he was all decked out in new overalls, noteworthy because, like most field biologists, he was not a great dresser. When we asked him what the occasion was, he explained that at the Wildlife Branch office he had seen a map with a line running through his study area. On asking what it was, he'd learned that it was the proposed gas pipeline. Even worse, an initial environmental screening had already deemed the impact of the line on the ecosystem he was studying to be acceptable. Realizing that his snowshoe hares would not agree, and that his plans to collect long-term data were at stake – hares have a ten-year population cycle – he came to town to address the hearing board.

We attended too, a mandatory sentence imposed by a summer contract with the Yukon Wildlife Branch to help it assess the pipeline's potential effect on caribou, moose, and wolves. Across the side wall of the gymnasium, below a faded 1950s portrait of the Queen, sat a line of newspaper people and assorted media. Across the front, under the basketball hoop, sat the august board. After a suitable introduction, up stood Charlie in his new overalls.

The address he delivered was even more stunning than his attire. Soon it was apparent that he intended to instruct the board that predicting any environmental future requires an intimate understanding of ecosystem processes that invariably rearrange and change things over time. He drew an analogy between an ecosystem and a bank. "To understand a bank – its stability, its resilience in times of trouble, its probability of persistence – you might superficially ask how much money is in it. Then you might ask how much, or what proportion of that money is in bonds, stocks, mutual funds, income funds, mortgages, venture capital, loans, debentures – the equivalent of 'species.'"

Already, the eyes of media people were beginning to glaze over.

"But no economist, no big-time investor, would be satisfied by just that," Charlie went on. "Economists want to know rates of change for each of these things, want to draw graphs of trends over time – a month, a year, a decade. Graphs to show trends are the bread-and-butter of all corporate reports and economic assessments. Investment counsellors can pop up endless graphs on their computers to show the track record of any economic enterprise – that is how they communicate, how they think."

The media people were staring off into space, suppressing yawns, doodling on their notepads.

"Even that is insufficient," Charlie persisted. "What adjusts those rates? How are they influenced by world events – a declaration of war, a mega-corporation fraud, a president's head cold? Are the answers wild guesses or do they have statistical validity based on substantial pre-existing evidence? Or, do events interplay with such complexity that no two situations are the same? What is the relationship between bond interest rates and shifts in the value of stock market shares, and why, and is the relationship consistent? What conditions precipitate runaway inflation or catastrophic decline?"

By then, the media people were slumping in their chairs, barely awake.

"An ecosystem is like a bank. If you want to know its stability, its resilience in times of trouble, its probability of persistence, you might ask how rich it is in species. But that is superficial information. What you really need to know is what species are present. For at least the key ones, those that are most abundant, rarest, major herbivores, summit predators, or keystone species variously defined as ecologically important, you need to know how many there are.

"Even that is not enough. Just like economists, you need to be able to make graphs, show change over time – in this case change in population sizes – and to understand the relative influence of birth rates, death rates, immigration, and emigration. Then, like the economist, you need to know why. How are these rates affected by range conditions, predation, competition, disease, weather? What about critical habitat requirements, migration, calving sites, winter range?"

The message was clear. Trends in species and rates of change and why they occur are crucial to understanding an ecosystem, and are a painfully obvious precursor to predicting the environmental impacts of any development. An ecologist deals with as much – even more – complexity as an economist. For example, an ecologist must consider spatial distributions – maps – a dimension that is irrelevant to an economist. Yet we undertake megaprojects with potential mega environmental consequences with as much understanding as a child with a piggy bank trying to figure out which income fund is most secure.

"The antidote is to upgrade our level of ecological understanding massively and quickly, but 'quickly' requires time to train ecologists and conduct long-term research that can show and interpret trends. Doing this is important, you see, because 'nature bats last.'"

The media did not catch on. They were asleep. Unfortunately, though less obvious, so was the board. You do not let some overall-clad university professor torpedo the preordained outcome, convince you to reconsider the script, undermine your determination to conclude what you are expected to conclude.

Through the whole proceedings, only one incident roused people. The issue, ironically, had nothing to do with the pipeline. Instead, it was second-hand smoke. Many attendees smoked during the proceedings, and by mid afternoon the picture of the Queen had disappeared in haze. Somebody asked the chairman to ban smoking. He refused. The next day a big banner was strung over the gymnasium door, reading: "Warning! Attending An

Environmental Hearing is Dangerous to Human Health." In a bold move, the chairman banned smoking.

The Alaska Highway Gas Pipeline was never built, but in keeping with what is now a long-standing tradition in Canadian environmental policy, it was not environmental impacts that killed the proposal. It was economics.

Shortly after the Yukon pipeline debacle, we started a decade-long "Yukon's Environmentally Significant Areas Research" program, which gave us new opportunities to think about ecosystem processes. We teamed up with colleague Gordon Nelson, whose expertise was, and is, landscape change. Our aim was to evaluate some high-ranking wilderness areas in the Yukon for protection as national parks or other types of reserves, and thereby act as catalysts for government action. We reviewed various approaches to evaluating wild lands, then cobbled together their strengths, stripped away their weaknesses, and, with some talented graduate students, invented the "ABC Method of Ecological Reconnaissance" – A for abiotic, or non-living; B for biotic; C for cultural. All three aspects were aimed at identifying and mapping processes taking place on the land and thereby recognizing areas of high ecological significance.

We identified four superb wilderness places with exceptional ecological and wildlife values and mounted field programs to evaluate them. As basic units for analyses, we divided each place into "biotic land units," usually a particular type of forest, or a grassland or bog whose uniformity we could discern from aerial photos. Then with our field crews we flew in by helicopter and spent weeks evaluating and ranking a representative set of these units. In the process, we saw and experienced some of the best wilderness in North America: spruce forests growing on top of a glacier; sand dunes being pioneered by arctic plants; rare species, such as the northern, disconnected population of Brewer's sparrows; escape terrain of Dall's sheep; and the calving grounds of caribou. We worked our way, quietly, around foraging grizzlies, and roped together to ford rivers with currents so strong they were rolling rocks on their streambed. We filled data sheets with information and notebooks with experiences.

Sitting on a rock knoll high up on a mountain or surrounded by an infinity of tundra, we discussed how to come to grips with what was going on. Where could you start? Could we develop an easy template for thinking about an ecosystem that anybody could use?

Our starting place was photosynthesis, the primary source of biological productivity. We asked what were the dominant species of vegetation that deliver it – spruce trees, willows, dwarf birches, sedges, or grasses. And why. The "why" is where process comes in. Were these species prevalent due to abiotic influences, such as the type of soil, or its moisture, or slope or aspect, or microclimate, or the influence of a disturbance like fire? Or, were biotic influences, like selective grazing or insect defoliants, the primary cause? To help figure these things out, we mapped where on the landscape these processes were most obvious or were likely to be influential.

We layered on other processes as crew members came up with them, mapping them where we could. Landscape ecology directs you to consider mosaics of ecosystems across broad landscapes by identifying the dominant vegetation types as a "matrix," and other places as "patches," and connections between them as "corridors." The next step is to consider what is happening between adjacent landscape units, such as the movements of nutrients through erosion, or transfer of water, or migration of wildlife. We then mapped all the key places.

Another way of thinking about and mapping process was to begin at the top, with the summit predators – wolf, coyote, lynx, wolverine, eagle – and ask where are their key ranges and critical habitats. Then, the same question for their principal prey species. Then, the same question for key vegetation food species.

We toyed around, too, with a layer that dealt with seasonal change: leaf fall and its influence on nutrient return; dormancy and its shutdown requirements, as well as the converse in the spring; life-cycle components like hibernation and migration. That layer did not work out, however, because we found that those features had either been considered already or were impossible to map.

Besides, we were soon overwhelmed by maps – maps that showed biological hotspots like wetlands; or saltlicks where ungulates come for minerals, especially in the spring; or south-facing grassland slopes that provide critical winter habitat for Dall's sheep and denning sites for foxes; or places where soil creep was occurring on steep tundra slopes underlain with permafrost; or where shoreline erosion and deposition were happening. These were the As and Bs of our study. From a cultural standpoint, Gordon's crew mapped the Cs: the traditional trails, hunting areas, and spiritual places of the local aboriginal peoples, and any nodes and corridors of early travel or settlement by first European explorers, miners, hunters, and settlers. Finally, we overlaid all these maps to identify the most significant areas for protection and wrapped them up with proposed boundaries for prospective parks or reserves.

We published the results in a series of reports, each with a photograph of Yukon's attractive Mount Hodge on the cover.[3] Then we waited for governments to come knocking at our door, eager to capitalize on our analyses. Alas, both the federal and Yukon governments were reticent to set aside lands for protection until the resolution of native land claims. A pipeline before settlement of claims? Of course. But a protected area? No.

Over time, however, various aboriginal groups have settled their land claims, and the Yukon government has initiated a system of territorial parks. Various agencies have stumbled across our reports, and citizen conservation organizations have used them as a basis for their proposals. But the full payoff of our wilderness analyses has been protracted, suffering from government inertia in land use planning.

Only a handful of studies have tried to go further than we did and identify everything that is hooked to everything else in an ecosystem. In the 1970s and 1980s, the International Biological Program attempted that on a few sites in Canada, including one on the prairies, another on a small lake near Vancouver, and one in the high Arctic. Flow charts attempted to illustrate the apportioning of energy as it moves from one species to another, or one place to another; other charts depicted the movements of various elements; yet others showed the interrelationships relevant to population limits of various key species. While the picture that emerged was of mind-boggling complexity, it was a pioneer attempt at a coordinated approach to understanding ecosystems that, if refined in the future, could yield insights.

Instead of coordinated research such as the work conducted by the International Biological Program, ecologists usually become specialists in certain species, or groups of similar species, or processes, and head off on their own. Sometimes they congregate at research stations, for instance, at the Wildlife Research Station in Algonquin Park, where we conducted both grouse and some of our early wolf research. For more than fifty years that station has catered to scientists engaged in field studies, which have provided a composite picture of the region's ecology, although that picture must be gleaned from a wide array of scientific and popular publications. More commonly, however, researchers pursue their own questions at whatever place is most expedient and least expensive. For that reason, ecosystem research has suffered.

But now we have an unprecedented opportunity, found almost everywhere – an experimental situation to die for, which many people will. Sudden climate change has the potential to reveal what are normally hidden ecosystem processes and relationships, an excellent research opportunity. What if today we had unlimited grant money and a large field force? We could envision setting up hosts of intriguing hypotheses – educated guesses to prove right or wrong. We would select as our study region the southern portion of the Canadian Shield in Ontario in the vicinity of Algonquin Park, because we know that country well. We would assume little change in precipitation – a weak assumption, but that is the best guess of climatologists for that part of North America, and making accurate ones for precipitation anywhere have proven difficult.

Being in an intermediate area between southern hardwoods and northern boreal forests, an easy prediction would be that with warmer temperatures the hardwoods would have little trouble becoming more dominant. All they would have to do is march off the rolling slopes into the lowlands where conifers now grow – a short-range, local expansion only.

Yet, on second thought, conifers grow in the lowlands not only because of cooler temperatures there but because of moister soils. After the conifers died due to heat stress, would overly moist soils complicate hardwood colonization? And what about soil acidity? The lowland soils are acidic – podzols – whereas hardwoods favour more alkaline conditions – luvisols. That difference would slow nutrient uptake and, in so doing, slow the basic rate of energy capture (productivity) as well, two vital processes.

On reflection, maybe the transition of the lowlands to southern hardwoods might not be so smooth. When the heat-stressed conifers died out, perhaps understorey hardwood shrubs, already there, would take over for a while and give the hardwood trees a short, or long, competitive struggle. And so, setting hypotheses for the effects of climate change on shifting forest types is not as easy as it first looks.

What about wildlife? Bark beetles and other invertebrates that spend part of their life cycles in dead or dying trees would flourish, as would woodpeckers that prey on these insects. That might favour the forest accipiter hawks, which prey on woodpeckers. But accipiters are generalist predators of songbirds, and still another hypothesis could be that the hardwood-loving bird species would decline, and the decrease in productivity would reverberate up the food chain.

However, that train of hypotheses may not be correct. Weakened and dying conifers are fodder for spruce budworm, which is known to increase the abundance of many warbler species. But maybe the accipiter populations would not track this increase in woodpeckers and warblers because something else is limiting their population, such as city lights attracting and killing them on migration, or loss of wintering habitat farther south. For insects and birds, then, competing possibilities make a best guess impossible. There are too many potentially significant environmental variables.

Then there is the large-mammal system. Surely we can make some accurate educated guesses about that. With more hardwoods, moose, we think, would decrease. Moose need conifer cover in winter and seek either lowland conifers or hemlock-covered ridgetops. If snow depths remain as they are now, or increase, without these conifer places moose likely would decline. Also working against moose would be a decline in balsam fir, a favourite food. A triple blow might come from the white-tailed deer population that could increase due to more hardwoods and bring in brainworm, a nematode that is lethal to moose (see Chapter 4). Finally, if deer numbers increased, so might wolves. More wolves could have a secondary effect of increased predation on moose.

However, deer might not increase, limited by human hunting or some other cause. So there might be no bringing in of brainworm and no increase in wolves. And a decrease in balsam fir as moose food might be offset by an increase in beaked hazel shrubs or red maple saplings, which moose also eat. Hazel shrubs, in fact, might become dominant if hardwood trees have difficulty recolonizing the lowlands. If snowfall decreased, winter cover would be less important. Once again, it's hard to choose among hypotheses.

Beaver would be glad to see the disappearance of the lowland conifers, which they rarely use, especially if the conifers were replaced by aspen or red maple, which are staple foods and also their materials of choice for engineering projects. More beaver, more ponds. Greater aquatic productivity. Frog populations would likely increase, but maybe that would spell their demise – greater density could cause an outbreak of the fungal disease chytridiomycosis that is currently sweeping the world and decimating amphibian populations.[4] Disease is density dependent. More frogs, or fewer frogs, could adjust the heron, bittern, mink, and otter populations either up or down.

This is a fruitless exercise. So many interacting processes, variables, and species are playing the game of life in any ecosystem that the complexity boggles the imagination and thwarts any attempts at prediction. Not even computer models can predict from the infinity of possibilities. We cannot even set up reasonable hypotheses. With climate change, we will just have to take whatever the altered ecosystem processes deliver.

How do you deal with bewildering complexity? With humility, or with prevailing arrogant, slapdash environmental impact assessments and an attitude of command-and-control?

CHAPTER TWELVE

TEAM PLAY

A tropical rainforest, rich and resplendent in biodiversity; a spruce forest, modest in species but vast in extent; the southwestern desert where life is spread paper-thin. Spruce bog, rotting log, an epiphytic bromeliad cradling its pool of microscopic life. All are ecosystems at various scales. All are live circuits composed of species, wired together by interrelationships, energized by the sun. What, as a team, do the living things in ecosystems do?

"Ecosystem" was voted the number-one concept – most important – out of fifty ecological concepts proposed and voted on by the membership of the British Ecological Society.[1] It did so by a wide margin, scoring 5.2 out of a possible 10, when its closest competitor scored only 3. Why did it rank above such vital topics as energy flow, carrying capacity, and co-evolution? How did it outdistance species diversity, competition, succession?

Some ecologists likely ranked ecosystem first because the concept, if understood by a wider populace, would reconfigure our relationship with the living world. Most put it first because it embodies intriguing scientific ideas that have been pursued since first articulated by the biologist A.G. Tansley in 1935. Every time you think you are getting close to a more complete understanding of ecosystem, like any good scientific concept it energizes itself with more subtleties and nuances and sprints away.

Most challenging about the concept of ecosystem has been the question of what, if anything, the amalgam of ecological processes, taken in concert, do? Individual plants photosynthesize, metabolize, respire, transpire, reproduce, and die. Individual animals eat, sleep, breed, hunt, hide, and die. But an ecosystem, a collection of species, just seems to exist, forever alive but accomplishing nothing – aimless, without purpose. And so, ecologists have been tantalized by a paradox: Are ecosystems simply the sum of their individual species, or are they more than the sum?[2]

When you look at an individual tree, bird, or flower, it just exists, with no concern for anything else. And another tree, bird, or flower, each maximizing its own welfare. The impression is that an ecosystem is simply the sum of its individual parts. However, especially over the past three decades, support has grown for the opposite view, that an ecosystem is more than the sum of its parts.

How can such a counterintuitive notion be valid? Only if one plus one equals something more than two, and in an ecosystem, that is true in many ways, any time two or more individuals form any sort of accommodation or alliance that produces something new. Examples range from a mated pair producing offspring, to herds or flocks exhibiting coordinated movement patterns, to species acting together creating niches or habitats for others to exploit. Even the antagonistic associations found in competition result in adaptations that can generate new species. And, in a most fundamental way, every individual both exploits and is exploited, making tiny to massive alterations in ecosystems.

Six Clark's nutcrackers – big, noisy grey-and-black birds – are swooping and diving among the ponderosa pines around me, calling incessantly with a rich vocabulary of squawks. The adult pair is busy feeding fully fledged young, who are drawing attention to themselves as they try their wings. They represent a new social unit, soon to form a flock and act as "ecosystem engineers," spending their days planting pine seeds. The seeds they plant but fail to find and consume later will form new pine groves,[3] where future nuthatches will probe the bark, pygmy and saw-whet owls will nest, crossbills and Steller's jays harvest nuts, larval pine beetles dig tunnels, and white pine butterflies lay their eggs. The sum of the effects of these six birds will spin on and on.

For ecosystems to be more than the sum of their parts, they must be rich in relationships, and they fulfill that requirement with ease: trophic relationships, predator-prey relationships, disease-host relationships . . . As well, there are many interlocking behaviours, such as mutualism – the reciprocal benefits two species can have on each other. An extreme of mutualism is shown by lichens, which join algae and fungi in a single organism, or less intimately by insect pollinators and flowers, or seed-dispersing birds and fruit.[4] What about bizarre relationships like slavery, shown not only by humans but by ants with their aphid livestock, or deception such as mimicry, shown when one butterfly species fools a predator by sporting a wing pattern almost identical to a poisonous species, or when a moth exhibits owl-like eyes on its hind wings? The rich variety of relationships imbue ecosystems with a host of novel attributes.

An ecosystem seen as more than the sum of its parts has been compared to a machine. Species play specific roles, like spark plugs, wheels, and driveshaft in an automobile, but, because of the way in which the machine is put together, unique properties emerge. For an automobile, the emergent property of the coordinated sum of its parts is linear motion. Lay all the pieces of an automobile on the machine shop floor and they do nothing. Only when organized in a certain way do they function. Only then does an emergent property emerge. Computers store and analyze information. A factory produces furniture, clothes, shoes.

Similarly, a sports team is more than the sum of the skills of its individual players. One of the prized possessions in our family is a hockey stick signed by the 1956 Montreal Canadiens, maybe the greatest hockey team in history. Great players were on that team: Jean Béliveau, Bernie Geoffrion, Dickie Moore. They formed a line, and their skill in playmaking as they swept into the opposition's zone was a marvel of team play. Surely their skill as a line was more than the sum of their individual talents. Like different components of an automobile, a hockey team has a division of labour, different tasks for different positions. The emergent property of the team, working together, is winning.

The question of what an ecosystem produces boils down to what its emergent properties are. Emergent properties have been described at various levels of the biological hierarchy, which, in order of increasing complexity, are protoplasm, DNA, cell, tissue, organ, individual, population, community, ecosystem, landscape, and biosphere. Various properties emerge at each level due to increases in organization. A population has a density, something that the next lower level, individual, cannot have. A community has species diversity, something that a population cannot have.

But ecosystems? In scientific literature, there is a strange absence of suggested emergent properties, save one – heat. Scientists have found that the more interconnections and complexity exist in ecosystems, the more rapidly they degrade energy – that is, the higher the proportion of incoming solar radiation reflected back into the atmosphere as heat. This discovery was made by flying over and comparing tropical forests and shrub deserts using suitable equipment to measure heat reflectance from the ecosystem below.[5]

It seems illogical, however, to think that ecosystems are the way they are, that they have evolved the complexity they have, just to be good at degrading energy to heat. Degradation may just be a by-product of ecosystems, like exhaust from an automobile.

Why emergent properties of ecosystems are so poorly understood is because the great bulk of research in ecology has focused on the isolation of individual parts, that is, individual species – studies of population dynamics, or evolutionary adaptations, or environmental relationships. This practice is like attempting to figure out how an automobile functions by examining each auto part separately rather than trying to fit them together in a functional way. Until recently, ecologists have tried to isolate, identify, and parse ecosystems down to their fundamental units. Doing so fits well with the scientific method because it limits confusing environmental variables. Identify a problem, such as how a population regulates itself, pick a species to study, pick a study area that is typical of a larger area, and go at it. Communities of species, in contrast, are much more difficult to study. And then to draw broad conclusions about what ecosystems do? That is a challenge.

For a long time, the question of what ecosystems do haunted our field work as we hiked or paddled along or sat by the campfire.

A little devil seemed to be sitting on our shoulders, asking, "Can't you figure it out? Can't you figure it out?" Is the ecosystem out there beyond the firelight just getting along, surviving the night, the season, the decade, century, millennium, and on? Is the game of life being played by plants, herbivores, predators, parasites, detritivores with no winner, no Stanley Cup, with only a whistle to end the contest when we are hit by an asteroid or run out of carbon dioxide (Chapter 15)? Or are we either too short-lived or too much a part of the biosphere ourselves to catch on? Worse, are there no emergent properties, and is the question that has nagged at us for so long simply fallacious?

But now, with immense satisfaction, we have finally figured it out. The little devil has gone. It was so obvious all the time! It was staring us in the face, but we had blinkers on. They were the blinkers of the reductionist science we had been taught, and the emphasis in biology on the lower levels of the hierarchy, and, curiously, the conventional understanding of the relationship between DNA and the cell.

The cell is thought to possess one truly remarkable emergent property – life itself. Inherent in any definition of life is reproduction. DNA reproduces itself only when inside a cell. Biochemists can isolate, even synthesize, a DNA molecule, but it only lies there in its flask or dish or whatever they make it in. For reasons yet unknown, it reproduces only when put into a cell. As well, cells metabolize, that is, capture energy and grow, a characteristic that defines life even better. Metabolism is the biochemistry of life. After all, crystals reproduce, but they do not metabolize.

Having long ago learned that life is an emergent property of a cell, it came as a shock to us to realize suddenly that it is not really true. That revelation came in a strange place, on an airplane, about nine thousand metres up (thirty thousand feet), midway between Toronto and Montreal. The artificial and imported ecosystem on the airplane, however, helped deliver the breakthrough.

All of us on Air Canada Flight 850 were the product of ecosystems, and more broadly the biosphere that was just then temporarily beyond us. We were breathing plant-produced oxygen and consuming ecosystem-produced tea, orange juice, flaccid fish, or rubbery chicken. Hydrocarbon products of ancient, decayed plants fuelled the airplane. We could not exist up there without ecosystem. Nothing living could, not even a single cell. And that was the breakthrough. A cell cannot reproduce or metabolize all by itself. It requires an ecosystem. Even the first cells on Earth could not have lived without nutrients and water – essential components of ecosystems. From then on, life has required ecosystems to create niches, impose competition, forge co-operation, shape natural selection, and in a host of ways provide all the services that configure living things. Every species in an ecosystem alters conditions that in turn influence the well-being of some or many other species. The first law of ecology, as expressed succinctly by biologist Garrett Hardin, is: "Everything is connected to everything else."[6]

At last we had the answer. Life is the emergent property of ecosystems! That is what ecosystems do – create, shape, and sustain life, turn the inanimate into the animate, breathe life into raw elements. Ashes to individuals. Life is their product. Life is the marvellous fabrication of the ecosystem machine, the ecosystem team.

So simple. Even a child with a cricket in a jar knows it has to provide it with an ecosystem to survive. But a child doesn't think of it that way. It just punches some holes in the lid to let in air and stuffs in some grass or leaves for the insect to eat.

But then the little devil came back, hopping up and down on our shoulders, asking what we, at first, thought was a stupid question, then realized was profound. "If a cell requires an ecosystem to become alive, but an ecosystem does not exist without cellular life, which came first, the cell or the ecosystem?"

Well, on reflection, the imp had a point, but only in a technical way, and only at the very beginning of life. Stripped to its essentials, life arose when early atmospheric gases interacted with land- and water-borne molecules through their chemical and physical properties. These interactions formed both RNA (ribonucleic acid, essentially a single strand of DNA) and a primitive cellular membrane that provided the RNA with its own immediate environment.[7] The very first cell obviously had to come before ecosystems with life in them, and it likely happened independently more than once. But, as paleontologist Andrew Knoll explains, "Once genes, proteins, and membranes were in place, life probably climbed the trunk of Darwin's great Tree of Life quickly, propelled by natural selection, gene duplication and gene transfer." As soon as cells, or cell parts like RNA, began to interact, as they had to, the ecosystem was on its way.

So profound were these interactions – cells with cells, cells with their environment – reproducing, differentiating, building reef-like stromatolites (sediments mixed with blue-green algae) that are visible even today, that within a billion years living things had totally overhauled the Earth. By then, ecosystems had created an oxygen-rich atmosphere. They had adjusted the levels of methane, sulphur, and carbon in water and air. Species had diversified enormously, with several major phyla of bacteria, a separate domain of Archaea (single-celled organisms with no nucleus, tough outer cell walls, and protective enzymes that allow them to thrive in extreme environments, among Earth's earliest life forms), and the first buds of our own domain, the eukaryotes. Living things, structured into ecosystems, evolved rapidly to assume planetary control.[8] In a way, life at first was like a hobo running for a train, swinging up into a boxcar. Once aboard, gaining sudden momentum, his legs became irrelevant. After that, the ecosystem train propelled life. And it has ever since, without pause, for more than 3.6 billion years.

This simple thought – the realization that life, in its wonderful complexity, is an emergent property of ecosystems – has been immensely satisfying, even when we just look out our living-room windows at the pygmy nuthatches and pine siskins in the saskatoon bushes, and the surrounding ponderosa forest, grassland, sagebrush, and lichen-covered rocks. The reward is a unity perspective, and we achieve it not only in our home ecosystem but every time we visit somewhere natural and new. The field approach to reap the reward is to avoid trying to achieve full-frame association with an animal or plant of interest, but rather to achieve a backed-off one that encompasses its sustaining surroundings. The accompanying mind exercise is envisioning species not just as entities, but as parts of a collective whole. Such a mental overview is facilitated by what philosopher and novelist John Cooper Powys called the "ichthian leap,"[9] metaphorically describing a fish leaping out of the water and, in that detached state, looking down and reflecting momentarily on the characteristics of its circumstances, its normal conditions of life, or here, its ecosystem.

The result is a sense of ecosystem completeness, of species combining with rich interrelationships to have a product – to sustain one another, and us, and the biosphere, in a mutual dance where the ecosystem, at once both crucible and content, breathes life into cells and species and populations and, in one glorious all-encompassing circle, back into itself.

The ecosystem as a team has another important characteristic besides collectively animating the Earth. While nothing can be as relevant as that, this other characteristic encompasses a particularly human way of thinking about nature. It relates to one of our most basic obsessions – health.

We knew Mount Allan in Kananaskis Country, Alberta, before its meteoric rise to fame. We understood it, too – its dignity of place among mostly smaller foothill mountains to the east, its moods in summer sun and winter storm, its pride in hosting what was probably the healthiest herd of bighorn sheep with the largest horns in the Rocky Mountains, its sorrow in hearing the howls of wolves only infrequently. We had taken its pulse on jubilant days, exploring its wild alpine gardens and frost-browned meadows. We had marvelled at huge conglomerate boulders, formed long ago at the bottom of an ancient sea and then hoisted high along the thrusting mountain folds. We had climbed its snowy flanks and cross-country skied the aptly named Ribbon Creek at its base.

Ten years before the eyes of the world fixed on Mount Allan, we had mapped its seasonal distribution of elk, white-tails, and mule deer both on the mountain and throughout the Kananaskis Valley. But our report, it seems, was stuffed away in some government office where it gathered dust. Even back then, rumours were afoot that things were going to change, that the mountain's open forests and gently sloping alluvial fans were attractive not only to the large mammals, which had shared them for eons, but also to planners intent on recreational development.

In those days, Mount Allan was in excellent physical condition.

A few circular clearings had been cut in the lodgepole pines on its southern face as part of an experiment to assess the effects of forestry on river flow, and most of the wolves that had once raced along its slopes had been killed or driven away, but otherwise, Mount Allan enjoyed fine health. Grizzly bears searched for hedysarum roots in the spring, and prairie falcons nested among the high crags.

A decade later we returned to the valley to discover that surveyors had been climbing Mount Allan. Armed with transit and chain and competence in cost-benefit analysis, they had mapped the gradient of potential ski runs, worked out the technical specs of making snow from river water piped up from below, and tallied the economic benefits the mountain could provide to Calgary and Alberta. Bulldozers were busy carving up the south face of the mountain; the simple little gravel road that formerly led to Ribbon Creek had vanished in a swirl of pavement. The asphalt sterility of gargantuan parking lots smeared the valley floor. The mountain – its alpine meadows, lodgepole forests, prairie openings, aspen-covered alluvial fans, bighorn sheep, and prairie falcons – had been placed on the altar of development. Mount Allan was to become the downhill-skiing site of the 1980 Winter Olympics.

Transforming Mount Allan into something "useful" represented no common enterprise. An environmental impact analysis was ordered to placate those environmentalists who were arguing for less ecologically damaging options. Boosters of environmental impact analyses claim that they can be useful in predicting the consequences of development and in minimizing the negative impacts. Critics say that these analyses result in no more than cosmetic surgery, and that they are often applied only as palliative care while the patient is being terminally bulldozed.

The medical analogy is more than a metaphor. Ecosystem

health is the subject of an increasing number of papers in ecological journals.[10] An ecologist's job is one of "ecosystem doctor": to take the pulse and blood pressure, to listen to the lungs, and to tap the knees of a landscape being subjected to development.

To think in such terms offers a constructive perspective on land and ecosystems. For example, a summary medical report on Olympic Mount Allan, ready for the big event, would have read something like this:

Name: Mount Allan.

Age: 75,000,000 years (when the Rocky Mountains arose).

Social Insurance Number: None. (Outside of national and some provincial parks and a few ecological reserves, Canada offers no security to land.)

Circulatory System: Hyperactive hydrological cycle caused by forest clearing and runoff from parking lots.

Metabolic Rate: Energy level low because of forest cutting and the impact of ski runs on alpine meadows.

Digestive System: Nutrient deficiency as a result of accelerated runoff causing slope erosion. Incomplete digestion from loss of large predators.

Respiratory System: Reduced plant respiration caused by vegetation loss.

Neuroendocrine System: Predator-deficiency neurosis – a pathological failure of feedback control on the large herbivores by predators, creating an imbalance in food webs that in turn reduces plant species diversity.

Pathogens: Suffering from acute attack of *Homo sapiens* on the south face. Scar tissue abundant. Healing will take a century even if the disease is cured.

Strengthening the parallel with medicine is the common denominator of stress, shared by humans and mountains. Just as Canadian pioneer stress researcher Hans Selye described the symptoms and consequences of excessive stress on human physiology,[11] ecological journals now describe the symptoms of stress on ecosystems. With humans, excessive stress may cause a failure of the body's hormonal and neural defence mechanisms. With ecosystems, excessive stress may cause a failure of the regulatory feedback loops. The consequences, however, are the same: departure from a set of norms that is interpreted as good health.

Ecosystem-stress research has resulted in the identification of several vital signs that scientists can measure. Just as a medical doctor diagnoses human stress from the body's efforts to cope with it, an ecosystem doctor looks for similar reactive signs. Stressed ecosystems, for example, exhibit changes in nutrient cycles and lose, or "leak," nutrients faster than healthy ones. Some human activities, such as logging, reduce plant cover; consequently, nutrients are not as efficiently retained. Conversely, various industrial activities create acid precipitation and air pollution, which typically kill bacteria, fungi, and other soil decomposers, leaving nutrients locked in dead plants and animals. Ecologists can assess both nutrient leakiness and reduced decomposition by measuring compounds in stream flow runoff.

A bedridden ecosystem may also exhibit lowered productivity, especially if vegetation has been killed or removed. And if production changes, so must, in a reverse way, respiration.

Furthermore, stress may result in a shift in the species composition of an ecosystem, a response analogous in stressed humans to the elevated production of such hormones as adrenalin. Typically, the losers are large-bodied, slow-reproducing species;

the winners are small-bodied, rapid-reproducing, hardy species, often non-natives that live in only small numbers until the environment is disturbed. With increased stress, even these replacement species may begin to disappear; terminally ill ecosystems exhibit chronic species impoverishment with violently oscillating numbers in the remaining species, often resulting in even more local extinctions.

Ecosystem medicine has some formidable obstacles to overcome, foremost being the problem of recognizing norms for any given ecosystem. The wide swings in what is normal cause interpretive difficulties, particularly because of natural disturbance and succession. As well, from a strictly scientific standpoint, many ecologists resist the temptation to think about an ecosystem as a single organism, partly because rarely do ecosystems have easily definable boundaries. Nonetheless, the value of thinking like an ecosystem doctor has more than paid off. It has provided a holistic way of looking at land. It binds together a bunch of processes that operate across a landscape into one image, one living thing, one supra-organism – something whose health we can care about.

Along with the rest of the world, we enjoyed the Winter Olympics. We watched the events on television, cheering the various victories. However, we could not help looking for the places along the ski runs where the elk had wintered and wondering where they had gone. We were disturbed when the television cameras panned across the surrounding landscape to show a snow-covered golf course where once we had watched foraging grizzlies. We remembered wild Mount Allan when it was just another mountain with only a benign level of human use, and to see it now, harnessed with permanent ski lifts and scarred from many wounds, conjured up the image of a sick patient in need of medical aid. It mixed our emotions.

On a humid summer evening, the heat pressed down on the tropical forest like a heavy hand. We were on the Indonesian island of Java, having arrived at the remote village of Panganderan the previous day via an overloaded ferry. From there we had walked along shrinking trails into the jungle. Here was one of the most impressive biotic teams assembled and currently active on the planet.

Our mission was to scout out research projects that could help establish a more sustainable future for the people of the Segara Anikan, a region on the island's south coast. Soil erosion following massive clear-cut logging was causing heavy silt loads in the rivers that discharged into the rich mangrove swamps of the broad Segara Anikan Delta. The silt was blanketing aquatic life and killing the fish that local people depended on for a living.

In the end, our mission failed, partly because the problem was intractable. The only solutions were depopulation and reforestation, and even then, revitalization of the rich marine life was doubtful. It was a cold lesson that recovery from environmental mistakes is not always possible. It was also a lesson that the levers for decision-making in some parts of the world are obscure, with power diffused through local, regional, and national governments in bewildering complexity.

But as discouraging as this realization was, our trip to Java had an enormous personal payoff. Near Panganderan we experienced, first-hand, a gold medal winner among ecosystems, still in reasonably good health. We were being guided through the rugged landscape of Taman National Park to a waterfall and the roosting site of huge bats known as flying foxes, which have metre-wide wingspans like that of an egret. On the way, we hoped to see a banteng (*Bos sondiacus*), a rare species of wild cow whose once wide-ranging stock has dwindled from several causes: hybridization with domestic cattle, domestication, hunting,[12] and a volcanic eruption in 1982 that spread a fine dusting of toxic ash on the grasses they ate. Only ten banteng were known to still live in the park. To help them survive, the trees in a few grassy openings where the cows graze had been deliberately hacked back to prevent forest ingrowth and habitat loss.

We threaded our way along a muddy jungle trail dwarfed by tropical trees of species we did not know, their limbs thickened with moss and draped with vines. The park plays host to more than five hundred native species of plants: palms, laurels, Diptocarps,[13] tree ferns, myrtles, and more. Two species of monkeys – one a highly endangered leaf monkey locally called "black monkey" (*Presbytis comata*), the other a more common and widespread crab-eating macaque (*Macaca fascicularis*) – appeared like escorts in the branches along the trail and boldly watched us pass, sometimes swinging parallel to us for a better view. Various jungle sounds accompanied us: the two-note, pigeon-like call of a tulung, the oriole-like song of a marcocu, the loud, bell-like call of a tropical kingfisher named cicak rowo, as well as the calls of many species of cicadas.

We reached a traditional banteng clearing where we found some fresh droppings, but their makers were sleeping off the afternoon somewhere else. As a consolation prize, we encountered a flock of Oriental Pied Hornbills – black-and-white, crow-sized birds featuring enormous, outlandish, yellow-and-white beaks with a huge top shelf called a "casque." Natural selection has had fun with hornbills, magnifying their calls with this resonating chamber.

As the afternoon wore on, we left the muddy trail to follow a creek bed, eventually gaining a view out over the ocean. We scrambled down to a waterfall, a thin stream dropping over a limestone cap rock, and there found the flying foxes *Pteropus vampyrus*, the largest of not only the fifty-nine species of flying foxes[14] but of any bat in the world. They flew sedately in a straight flight path, much like egrets, not erratically as bats usually do.

Rapidly, as is characteristic of the tropics, the evening light faded. The night shift of jungle sounds punched in, beginning with and then dominated by the shrill calls of frogs, including one that our guide informed us was a whopping thirty-eight centimetres (fifteen inches) long. The two species of monkeys still chattered at our intrusion. Soon, we were standing in darkness, and inopportunely our guide announced that he had brought no flashlight. Darkness on the floor of a tropical forest is rather complete, especially when storm clouds threaten, as they did that evening. Someone found a pencil flashlight in a side pocket of a day pack, but the weak beam was next to useless. Before we had gone far, it was dropped on a rock and went out. We groped our way along, each person unable to detect even the person just ahead.

Moments later, the storm struck, pelting the forest canopy and plastering our clothing to us. Our guide lit his cigarette lighter for whatever help that would be, but rain quickly doused the feeble flame. Making conditions even worse were rocks on the trail, undetected until we stumbled over them. We joined hands and felt our way along, the leader periodically calling out, "Root," or "Rock."

Eventually we made it out and back to the lighted streets of the village where we had booked a room. We attracted a few good-natured (we presumed) shouts from locals on their porches, who'd had enough sense to get out of the rain.

Lowland tropical forests such as this are among the most diverse and complex ecosystems on Earth. This example on Java, at only eight degrees south latitude, was the biologically richest place we had ever been. As frustrating as it was to be unable to identify even the common species around us, we sensed a deep satisfaction in being surrounded, encased, immersed in all that life, to experience what an ecosystem machine in excellent condition can achieve. The storm and darkness magnified that sense of immersion, of drowning in a sea of life. Without vision to distract us, it was easier to submit to the wholeness and completeness, the running together and mixing of the currents and undercurrents that entwine to create life.

A tropical rainforest, rich and resplendent in biodiversity; a spruce forest, modest in species but vast in extent; the southwestern desert where life is spread paper-thin . . . What do these ecosystems do? They animate the Earth with staggering biochemical and cellular activity. They ripple with conduits and links and relationships among species. The processes that structure them provide not only the foundries for life but overarching guidance for its self-organization.

PART 4:

PROSPECTS

HAVING TEASED OUT ANSWERS to the basic questions about
the self-organization of life, we can now apply them to future
prospects. Let's ignore the completely unforeseen, such as a
random hit from a giant extraterrestrial object that creates a prolonged nuclear winter or alters the planet's orbit. To see into the
future we can only extrapolate from the past. Representing different perspectives, first we examine the resilience of ecosystems, that
is, their ability to bounce back from whatever we, or nature, throw
at them; then, aspects of life's struggles to stay within the planetary "habitable zone"; and finally, how life handles hostile places
and what that means for the future.

CHAPTER THIRTEEN

OUT OF THE SLAG HEAP
OF NATURE

Most ecosystems on Earth are reduced, periodically, to rubble. Nature's ruthless wrecking ball swings down and smashes them. We humans manage to help with inappropriate indifference. But, in a seeming miracle of self-orchestrated engineering, ecosystems recover. Because of this ability, ecosystem resilience has received considerable attention. Yet how resilience is achieved remains contentious, its practical limitations unresolved, and its meaning for the future unclear.

ECOSYSTEM RESILIENCE IS A VITAL restringing of ecosystems, a rebuilding of structure and diversity and relationships. All but remote areas of the globe consist of human-altered, second-hand landscapes, or third- or fourth-hand . . . In the future we will ask for extensions to fifth-, sixth-, seventh-hand . . . and still expect these altered environments to deliver all the ecosystem services that are required for life on Earth. Recently we have added the expectation that ecosystems will recover from new insults, such as widely distributed toxins, abrupt human-caused climate change, genetically modified biota, and accelerated rates of species extinction. If we understood more clearly how resilience comes about and is maintained, we might face a brighter environmental future. Can we expect ecosystems to heal themselves forever, or might we eventually be called on to administer first aid? But how? What do we really know about ecosystem resilience?

The delivery boy for *long-term*, continent-wide resilience is natural selection. It took the Earth millions of years to recover from the meteor strike that ended the era of dinosaurs 65 million years ago. We only dimly perceive resilience at this magnitude, pieced together from scraps of evidence by paleontologists, paleobotanists, and other paleo-sleuths. Across all but the polar regions and parts of extreme eastern and western coastal North America, the meteor strike flattened forests, sterilized soils, and decimated species.[1] Yet moments later, geologically speaking – a few million years – lush forests grew almost everywhere once again, ecosystems hummed along once more, and species diversity had bounced back with a new cast emphasizing birds and mammals. That new cast required evolution, stimulated here by the opportunity presented by the vacant dinosaur niches. Similarly, it took millennia for ecosystems to rebuild after ice repeatedly blanketed the northern third of North America over the past 5 million years. However, rebuild they did. Life is tenacious.

One of the key sleights-of-hand of ecosystems in achieving long-term resilience is the ability to retool species to fit altered environments. Sometimes this process of speciation is achieved by extinction and, with time, refilling of the vacant niche by some other species that splits off as a deviant from an existing one and adjusts to the new role. Other times, the niche is never left completely vacant, and a gradual transformation takes places in the species occupying it, like an animation that begins with a child's face and renders it in its adult version. Whichever way speciation is accomplished, a long parade of different animals has marched in evolutionary progression across the world stage, driven by ecosystem processes that persist.

Periodically, major innovations have taken place. Primitive invertebrates joined the parade 800 million years ago, vertebrates 500 million years ago, insects 430 million years ago. Because of the spotty fossil record, we see the emergence of such major life forms as deceptively sudden. In fact, they occurred in stepwise progression, sometimes with unusual bursts of novelty but never as one-leap events. Today's species are marching in front of evolutionary lineages that can be traced back into the distant fog of time.

Striking patterns are apparent in this parade. Body forms repeat themselves. Near the front, for example, is a characteristic dog-like body plan – a wolf. Then, farther back is another beast with the same basic plan – a Mesonyx, a wolf-like predator that lived 45 million years ago,[2] closely followed by a Hesperocyon, an even earlier dog-like mammal. Much farther back, now in the Mesozoic 150 million years ago, with no direct linkage to these latter-day mammals, marches a mammal-like reptile with the same body plan again – a Dog-faced Cynodont, companion to the dinosaurs. Even farther back comes a dog-like member of the older Dicynodont reptiles.

Amazingly, all these species share a similar body form. All are medium-sized predators. All played roughly the same role in their respective ecosystems. Yet only the most recent three are related.

The older reptilian lineages of dog-like animals either came to a dead end or led to smaller, insectivorous early mammals that only later radiated into the modern orders like the Carnivora that exist today. But the body plan was so good that evolution repeated it again and again.

Similarly, you can trace back a repetition of body forms starting with almost any living thing today: birds through bats through pterosaur reptiles through ancient dragonflies with half-metre wing-spans. The same repetition occurs with large grazing animals, and with small mouse-like ones. Repeatedly through the ages, the various successful body plans have been constructed by a constant set of ecological processes. Life, when poured into the mould of ecosystems, structured as it is by natural selection and with a constant set of trophic, energy, and nutrient relationships, has turned out only slightly altered versions of a few body plans, time after time.

We sensed that great continuity one hot day in Wood Buffalo National Park on the Alberta – Northwest Territories border. After crossing the mighty Peace River in a motorboat, we began a sweaty trek to our campsite. Leader of our little expedition of five was Sebastian Oosenbrug, who was studying wolf-bison relationships. At the edge of a clearing we halted for a breather, easing our packs to the ground. Sunlight filtered through aspen leaves and dappled the grass. The sweet smell of aspens permeated the air, mixed with the fetid odour of bison dung. A cicada hummed in the trees, and every few minutes a yellow-rumped warbler sang languidly. We spotted it searching for and gleaning larvae from the undersides of leaves, fuelling up for the migration ahead.

Suddenly, on the far side of the clearing a shape emerged from the trees – a big bull bison. We were close enough to see its red-rimmed, beady eyes. It walked out into full view, head up, then turned to look around, searching for wolves or any movement that might mean danger. We stayed motionless. Satisfied that all was well, the bison began to graze.

Palaeontologists have a particular advantage over ecologists. They deal with the past, and consequently may have a heightened sense of time. No doubt they look up from their dig site, on occasion, and imagine the earlier scene. But an imaginary time warp is open to anyone, an indulgence that puts the here-and-now – and us – in perspective.

We fell into one such time warp on the edge of the clearing. The cicada droned on. In hushed tones, we speculated on what had once been there. The bison, now in our imaginations, transformed into a giant long-horned bison, a species extinct these past ten thousand years. The aspen trees melted away and were replaced by tundra. The warbler, too, was gone, and in its place was a tundra-dwelling yellow-wagtail, gleaning insects from the ground, fuelling up for the migration ahead. Still, the sunlight poured down, the fetid smell of dung clung to the air. The giant bison's head was up, eyes scanning the tundra, searching for sabre-toothed cats or dire wolves, alert to detect any movement that might mean danger. Satisfied that all was well, the giant bison began to graze.

Ten thousand years ago is just yesterday. What was in this place 30 million years ago? The vision of the giant bison grew even larger, its shoulders more humped. Its horns faded away and, instead, a rhinoceros-like tusk emerged on its forehead. It was recognizable – a Brontothere. Gradually, trees re-emerged behind it, some with aspen-like leaves on them, and the clearing reformed. We were in a slightly familiar temperate deciduous forest. The cicada was humming in the trees once again, and a small unrecognizable bird built very much like a warbler was gleaning insects from the leaves, fuelling up for its migration ahead.

The Brontothere slowly moved his huge head from side to side, searching for Andrewsarchus, as well it should, because it was the largest meat-eating land mammal of all and one of the Brontothere's only predators. Seeing no Andrewsarchus, the Brontothere walked over to what looked like a sycamore and began to browse.

What about 150 million years ago, in the age of dinosaurs? The Brontothere, large as it was, began to expand, to elongate, especially its neck, and it developed a long, whip-like tail. Spines developed along its back, its head shrank, its horn disappeared, and in a matter of seconds, it took on the unmistakable shape of a Jurassic Diplodocus. The sunlight began to fade and the temperate forest became denser, more tropical. Tree ferns sprang up around us, and giant redwoods, and cycad trees. The decidedly dank smell of wet, decaying vegetation reached our nostrils. A cicada-like insect still hummed in the trees, but this time no bird. Instead, a pterosaur, or a bird-like dinosaur called an Anurognathus, with a modest wing span of about half a metre, swooped back and forth in the branches, picking off insects.

The Diplodocus's tiny head, only a swelling at the end of its skinny neck, scanned the scene. It was searching for the dreaded Allosaurus, large enough to kill its young hidden in the patch of high horsetails. But the coast was clear, and slowly raising its head to the crown of a cycad it began to pull off its new leaves.

"Time to go," Sebastian called. "Another few kilometres to the campsite." The Diplodocus vanished.

Processes underpinned what species swirled out of cosmic dust in the past to march on the world stage. The same processes will fashion what comes next, and provide reasonable assurance that, on into the future, there will be a biological next.

In contrast to long-term resilience, the delivery boy for *short-term*, smaller-scale resilience is primarily succession, which is the universal healing process for disturbance both on land and in the water, and in microbial, plant, and animal communities, from polar regions to the tropics. Succession is the orderly replacement of one community by another, each adjusting the physical and biological conditions for the next in a marvel of uncoordinated planning. In discussing succession, because the living parts of ecosystems tend to drive change, the emphasis is on *community* (the living) rather than on *ecosystem* (the living plus inanimate), but in most contexts here, the two terms can be used interchangeably.

Exactly how succession delivers resilience is surprisingly contentious, despite decades of study and debate. That debate began in the 1920s and centred on two men who have gone down in posterity as "fathers of ecology." One of them, Frederick Clements, growing up on the Nebraska prairies, not only saw what everyone saw – how the land rebounded after devastation from plough, fires, tornadoes, and overgrazing – but delved into how it happened. He concluded that succession is driven by entire communities of species with environmental needs and demands that alter the environment, then are replaced successively by other communities with requirements better able to exploit the new environment. It was obvious to Clements that succession was driven by collective properties of species that emerged at various times.

———

Straddling the Oak Ridges Moraine in southern Ontario is the 250-square-kilometre watershed of the Ganaraska River. Here was our equivalent to Clements's Nebraska plains, our learning ground for succession. Once, the Ganaraska was a land of steep-sided sandhills, sculpted by glaciers and robed in a rich forest tapestry of maples, oaks, and giant white pines. Once, it was the fount of crystal streams cascading from secret, cedar-draped springs, where brook trout spawned and white-tailed deer came to drink. Once, it knew the waterfall-sound of millions of passenger pigeon wings. Once, the wind that stirred the leaves and wrinkled the surface of the scattered ponds was washed by wilderness. Back then, it was pristine.

Fourteen thousand years ago, during a time period called the Port Bruce Stadial, remnant glaciers lying in what is now Lake Ontario advanced north for about thirty kilometres (nineteen miles), ploughing up former lakebed sediments into sandhills and ridges.[3] Spanning all the subsequent thousands of years – the warm Hypsithermal period and the cooler Little Ice Ages – grasslands and forests came and went but constructed only a thin veneer of organic soil. Sandhills are inherently unstable, their soil nutrients in delicate balance with the overlying vegetation. That vegetation glues the fragile ecosystem together.

In this landscape, the deciduous component of the forest assured that all-consuming fires would have been rare. Every so often a ground fire would have licked the forest floor clean, but the effects of such fires become unnoticeable after a few years. Sometimes a windstorm would have knocked down trees, but that would only have ushered in more sunlight to release the cherry and ash seedlings patiently waiting their turn. The wandering habits of the Mississauga Indians, and before them the Neutrals, dictated that their occasional camps or clearings caused only ephemeral disturbances.

All that rule-by-nature changed in the early 1800s. Loggers came to those same sandy hills to cut the huge pines. Future masts for the Royal Navy were floated downriver to the newly minted town of Port Hope on Lake Ontario, where they were loaded onto ships bound for Britain. Massive, squared timbers and cut lumber were part of that export trade.

Only a few years later came the first homesteaders, with a mandate to clear their allotments or lose them. Surrounding those early settlers was an abundance of wildlife. The sandhills grew stands of beech trees that attracted passenger pigeons, whose flights at times "darkened the sky."[4] According to early records, waterfowl were "scarcely less conspicuous than pigeons in this early land of plenty." Salmon were superabundant, with stories of people spearing three hundred in a few hours. Bears, deer, grouse, and beaver were plentiful. Legend has it that wolves "stalked and howled around the settlers' cabins at night with great persistence."

With axe, oxen, sweat, and no doubt tears, the pioneers refitted the land to meet their needs At promising mill sites, settlements took root, with a store, livery stable, blacksmith's shop, church, and school. One typical settlement blossomed beside the Little Ganaraska River, Elizabethville, named in 1840 by the first postmaster for his wife. Though the town began with only four families, the entrepreneurial postmaster had soon constructed a combined sawmill and gristmill. Those were the key requirements to attract more settlers, to drive "the bush" even farther back, and to "improve the land," as early books called the settlement process. At Elizabethville, soon a church was built, and to round out the necessities of life, both a distillery and a tavern.[5]

For a time, the sandhills around Elizabethville were prosperous, with scattered farms spaced among the largely forested hills. For a while, it was a place of hard work, human happiness, and hope. The productivity of nature and of human endeavour joined in a degree of accommodation and balance.

Elizabethville was not alone. Other mill sites expanded to become settlements at Kendal, Garden Hill, Decker Hollow, Osaca, Knoxville, all of them mostly ghost towns today. Like raindrops rippling outward in a pond, each settlement was ringed with an expanding circle of cleared land. As the storm of development intensified, ring met ring until land transformation was complete. By then, the original fences of stone hauled laboriously from the fields served only to hem in the remaining patches of forest.

But, as often happens, humans went too far, failing to distinguish between living on nature's interest and squandering its capital. That distinction in rural landscapes occurs when forests become so fragmented that they lose their natural species composition, and wildlife is exploited beyond its annual productivity, and soils are robbed of their nutrients. Then, nature forecloses.

It foreclosed on the Ganaraska sandhills. The topsoil was too thin, the hills too steep. When too little forest remained, and consequently the natural regimes of wind and water were altered too far, then the soil eroded, the crops withered, and the livestock died. Rapid runoff gouged deep gullies in the barren slopes, streams carried away tonnes of soil, silt, and sand, and the salmon that had once spawned in the streams were smothered in the debris of collapsed ecosystems. Port Hope, at the river's mouth, suffered a long series of disastrous floods. Beaver were trapped out, deer overharvested, wolves, bears, and lynx exterminated.

Then, it was a place of heartache and dispossession. By 1885,

Elizabethville had begun to decline, and the other villages followed. Farms were abandoned, roofs caved in, and buildings disintegrated in patches of weeds and vines. Lilacs planted outside now grew inside, too. Nothing was left but a second-hand landscape of shifting sand – used, discarded, abandoned.

Robbed of its vitality, the land lay largely vacant for decades. The Dirty Thirties did not help. A government report on the neglected Ganaraska watershed of 1944 displayed the following quote on its frontispiece: "How can people do such things to their own country – weaken its base, befoul its beauty, darken its future? How can they countenance and join in a continual defacement and destruction of the body of their land?"[6]

Many such used, abused, and forgotten landscapes exist in eastern Canada and the United States, where scenarios of similar ecological misunderstanding have been played out. They are places of ghost towns. They are found anywhere perceptions of agricultural capability failed to match ecological reality. Lush-appearing forests do not always indicate potentially rich farmland, not where thin topsoil is underlain with sand or bedrock,

or where the growing season is short, or where the land is prone to flood or drought.

Yet, remarkably, the Ganaraska country recovered. Once the pillage ended and the marauders had departed, nature made a comeback. As well, recognizing the disastrous impact of settlement, and as retribution for past land-use sins, a conservation authority was established. Jobs were created, people began planting trees and building check-dams. Land on the upper part of the watershed was purchased as a conservation reserve. However, while humans provided some of the undercoat, it was nature that painted the picture.

I (John) knew the Ganaraska country as a youth. I tracked deer there, learning how they often circled and stayed in heavy cover when being pursued. I fished for brook trout, camped in the hills, counted grouse along the winding forest roads, and saw my first northern black-backed woodpecker.

Sometimes I came upon broken-down, crumbling foundations of old houses or barns deep in the woods. Occasionally I found and followed stone fences that ran through forest separating nothing from nothing, or I discovered old stump fences made from the roots of giant pines with sufficient pitch to prevent rotting. The pine plantations had matured, been thinned, and by then grew maples and beeches in the openings. While it was a second-hand landscape, it was lush with regrowth, wildlife, and adventure.

Sometimes succession takes mere decades, and the new landscape closely mirrors its predecessor. Those times are marked by two characteristics – soils left relatively intact, and nearby reservoirs of species. At Ganaraska, the effects of axe and plough applied repeatedly to its unstable soils left a largely barren substrate of sand dunes over much of the landscape. Had the disturbance been caused by natural events, such as a fire, hurricane, or insect outbreak, ecosystem recovery would have been quicker, because damage to the soils would have been slight. Moreover, natural disturbances often result in a sudden enrichment of soil by burned or decayed plants, firing up the rate of productivity above the norm.

Aiding recovery at Ganaraska, however, were remnant forest patches that served as reservoirs for dispersal of seeds, spores, and insects. Leaves would have blown from the forest and accumulated in sheltered places, beginning the process of soil development once again. The closer such sources of renewal, the faster recovery will happen.

Because our home town of Oshawa was nearby, now and then over the next thirty years we returned to walk those same deer trails and back roads. With time, the forest canopies became progressively denser, the road edges thickened with raspberry vines, the last remaining fields closed in, first with sumac, then maples and oaks. The shrubbery was enriched with hobblebush, viburnum, and beaked hazel. Mats of bearberry, goldthread, and twinflower expanded and invited in a variety of wildflowers – red columbine, moccasin flower, and trilliums aplenty.

Ganaraska country serves in our memory as a shrine to nature's resilience, its ability to bounce back, evidence of the restorative powers of ecosystems. We think of Ganaraska at times when we witness the too-heavy hand of humans on various landscapes. Doing so provides an antidote to the ecologists' burden of knowing what people are inflicting on the Earth.

The person arguing against Clements in the debate over how succession delivers resilience was botanist Henry Gleason. Although tutored in Clements's philosophy, he challenged Clements's concept of succession, arguing instead that succession is driven by individual species acting independently. His viewpoint fitted well with a reductionist way of looking at nature, specifically, that an entity like an ecosystem could be explained solely by examining its parts. He published his challenge in 1926, which Clements simply ignored. With uncharacteristic ambivalence, most ecology books of today simply present these two ideas as competing hypotheses without favouring either one.

This debate has left a mysterious aura around succession, a sense that some hidden alchemy is involved. Moreover, if Clements's view (community-driven succession) is correct, then it may require a rethink of a key premise of natural selection: that it works only on individuals. Maybe that is why succession was ranked the second most important concept in ecology, only after "ecosystem," by the British Ecological Society.[7]

Frank Miller's cabin was one of the original log roadhouses in Alaska, a picturesque, low, rambling structure rooted on a bank of Miller Creek. By the door lay the horns of a giant bison that some miner had unearthed from the creek decades earlier. A spur road ran down to the roadhouse through stunted spruce and birch from the gravel road called the Steese Highway.

Frank was in his nineties when we knew him. Throughout the early days of Alaskan gold mining, he had washed his share of gravel. He had been in Dawson in the Yukon Territory during the gold strike, then walked a few hundred kilometres to Nome, Alaska, then Fairbanks. His wife, Graziella, had lived in the gold camps, too. A seamstress, she was one of the first white women to brave the Alaskan frontier. Their stories were rich in humour and adventure. Frank was famous because he and his brother, "Cow" Miller, had driven the first cow across the Chilkoot Pass to Dawson and sold her milk to the saloons at thirty dollars a gallon.[8] On many evenings we tape-recorded old Frank recalling his historic past as he rocked back and forth in his ancient rocking chair before a Yukon stove made from a 200-litre (55-gallon) oil drum. The stove threw out so much heat that we had to move back halfway across the room, but old Frank sat there in his red flannel shirt and long underwear within centimetres of the heat, just pleasantly warm.

The first assault on Miller Creek came in 1894. Pickaxes, shovels, and wheelbarrows were used to move gravel to sluice boxes, where the debris was washed away and separated from the flakes of gold. The second attack, a few years later, involved more damaging "hydraulickin'," where big hoses blasted the soil and gravel off the creek benches and into the sluice boxes. A third invasion, mounted in the 1930s and 1940s, employed house-sized gold dredges that crawled along the creek bed reworking the tailings, picking away like scavengers at the corpses of gravel.

When we first saw Miller Creek, three decades later, the washed out gravels looked just as they had the day they were spewed out of the dredge. Rate of succession: zero. Ecosystem resilience: nonexistent. The gravels were demanding primary succession: succession on a substrate left barren of organic matter. Over the years, some alder leaves had drifted around the base of the piles, beginning, as they decayed, the process of rebuilding soil. Some fireweed and grasses grew there, too.

However, for soil to develop higher on the piles was going to be painfully slow. First, it would require lichens, unique among plants in being able to obtain nutrients from the air. Then would come mosses, plants without roots that can pioneer on the thinnest of lichen-built soil. Both of them would invade with spores blown from the slopes above. Next would follow shallow-rooted flowering plants with light, wind-blown seeds. In the north, fireweed, the floral emblem of the Yukon, is the archetypical pioneer.

Grasses with surface-spreading rhizomes often accompany fireweed. As each year's growth decays, the soil deepens and, in an inappropriate metaphor, paves the way for the first shrubs. Alders come in early, with nitrogen-fixing bacteria in their root nodules that convert atmospheric nitrogen to usable nitrate, instead of relying on its uptake from the soil. Willows and birches arrive as shrubs, different from their related tree species, with the capacity for growing lateral roots in shallow soils.

Because of permafrost, that is as far as succession could go in the shrub-tundra on the lower slopes above Miller Creek. However, in more sheltered draws or warmer south-facing slopes, succession would move on to sun-loving white birch and aspen, shading and thereby weakening the pioneering ground flora. Finally, in places left alone for a hundred years or so, shade-tolerant white spruces would invade, eventually topping out the earlier trees and declaring dominance.

The "beauty" of Miller Creek was that all these stages of succession were on display simultaneously. The length of time since the different assaults of the miners, along with their varied impact on the soils, had allowed succession to proceed to different stages. Situated above the dredge piles, with their primary succession just beginning, were the worked-over benches of the "hydraulickers." There, organic soils had either been left or had washed down from above, so that secondary succession was in progress – succession which, at Miller Creek, skipped the necessity of lichens and jumped in with mosses, flowering plants, grasses, and, where old enough, dwarf birch and willow. Yet other places had escaped "hydraulicking" and been subjected only to the pickaxe and shovel of early days, and the years had reconstructed the forest of spruce, birch, and poplar once again. Moose and caribou browsed there, gray jays sifted through the tree branches, and golden-crowned sparrows and varied thrushes piped their lonely notes once more.

Frank died in 1969, and only months later, so did Graziella. For years before their deaths, officials at the University of Alaska in Fairbanks had entreated the Millers to donate their historic roadhouse to them, but the Millers never understood why – it was their home. Instead, they willed it to their friend, a German immigrant who had been a doctor in the old country but never was licensed in Alaska. "Doc" drove the grader on the Steese Highway with his doctor's bag stashed near the clutch, and administered to the people living in the hills, missing licence notwithstanding. Sadly, within months, he was killed in an accident on the highway. His estate sold the roadhouse to miners who knew that the roadhouse stood on virgin ground. The Millers had regaled too many people with stories about finding flakes of gold from their well in their drinking water. Predictably, within months the roadhouse burned down. The site was bulldozed, mined, and abandoned.

Twelve years later, we were in Fairbanks and decided to go out to Miller Creek and relive a few memories. We drove down the spur road, surprisingly straightened and widened, and to our dismay, instead of seeing the roadhouse, saw fireweed and grass. Along the creek bank, struggling wild rose bushes and silverberry were eking out a tenuous existence, and, ironically, some mountain forget-me-nots. Then, in the vegetation, we saw the old Yukon stove, rusted, lying on its side, the door gone – the same stove we had sat beside so many times while Frank spun his yarns. One aspect of succession that the ecology books never mention is the human memories it obliterates. Maybe, someday, succession will have the last word and obliterate the memory of humans on this planet altogether.

The orderly progression of events during succession at both Ganaraska and Miller Creek seems to beg for an explanation that goes beyond the uncoordinated activities of individual species. Each stage, from pioneer to mature community, is not only comprised of a set of interacting species with characteristics suitable for the conditions, it also prepares the stage for the next set of species. In the progression, short-lived, light-demanding species give way to long-lived, shade-tolerant ones.

The more succession, as an overall process, is driven by sequences of events such as changes in shading, the more orderly it will be. Several sequential events play dominant roles. In early stages of succession, energy capture and productivity (fixation of green material) is so high that plant material accumulates faster than it can decay. As succession proceeds, the amount of green material (chloroplasts) rapidly increases, and so does the total weight of both living and dead material. However, shading from a forest rebuilding vertically – shrubs, saplings, trees – gradually puts a brake on productivity. The rate of photosynthesis per chloroplast declines, because many leaves are no longer in full sunlight. In late succession, productivity balances respiration, and plant material reaches a stable equilibrium with decay.

Another sequential event, especially important for resilience, is an increase in the number of species. Rural southern Ontario, for example, consists of a rolling agricultural landscape with a patchwork of woodlots in various stages of succession following logging – logging that includes a range of practices, from carefully managed to brutal. In a recently razed woodlot, only a few sapling willows and aspens poke through raspberry vines and red elder bushes to start succession off. Slightly older woodlots feature a scattering of hawthorns and buckthorns as well, and a few struggling black cherry trees. Then, in come some white ash, and maybe, on warmer sites, a few red oaks, and when the site is older, some shagbark and bitternut hickories. As the shade intensifies, conditions become suitable for sugar maples and American beech trees, species that propagate best in mature forests.

Thus, species diversity increases, at least until the very end, when a completely closed canopy ushers out the sun-requiring species. Driving this increase is a greater variety of habitats and niches as the community becomes more complex, particularly as it expands vertically. But what matters most to the community is the resulting increase in the number of alternative pathways for energy and nutrients to follow. More pathways means a greater ability to bounce back, but also greater resistance – and ability to minimize disturbance in the first place.

Illustrating the considerable value of alternative pathways, two tree pests are slicing through southern Ontario. One, the Asian Emerald ash borer, kills white ash. The other – an introduced scale insect whose damage allows several fungal diseases to develop – kills beech. Similarly, decades ago, a bark beetle brought Dutch elm disease to American elms, wiping out their former abundance in the eastern forests. Yet, as there were alternative pathways for energy and nutrients, all that happened was a shift in relative dominance of the remaining tree species. The forests quickly recovered. In contrast, in the early 2000s, when mountain pine beetle hit single-species forests of lodgepole pine in British Columbia, energy and nutrient pathways failed as vast tracts of land became devoid of live trees.

Yet another sequential event taking place during succession is a shift in the character of the species involved. Species of early succession are the pioneers – the ecological band-aid species, the colonizers, the risk-takers. They tend to be opportunists, are generally small, grow rapidly, have short life cycles, and produce plenty of offspring. They are ecological generalists, able to tolerate fluctuating environments[9] and capable of dispersing widely into a broad assortment of new habitats.

But progressively, through succession, physical conditions become more stable, and selection pressure shifts to a greater emphasis on competition. The result is habitat and niche specialization. As well, species tend to be larger, longer lived, produce less young, be less dispersive, and either hoard resources or use them more effectively.

Textbooks describe these two different suites of characteristics, for generalists and specialists, as occurring in neat packages – maybe too neat. During a sabbatical in New Zealand, we realized that its introduced herbivorous mammals, of which there are many, provided an excellent opportunity to test this contention. Without the depressing or confining influence of predators, the colonizing herbivores were free to exhibit their natural or potential rates of productivity, dispersal, and habitat choice.

New Zealand is a disturbingly beautiful country – beautiful for anyone who likes wide-open vistas, precipitous mountains, and open plains; disturbing for ecologists who, by virtue of their profession, can see the screwed-up ecological systems behind the scenery. When New Zealand drifted off into the South Pacific between 85 and 70 million years ago, it carried no mammals with it. The reason is obscure, because marsupial mammals, and even small, insectivore placental mammals, had lived on its mother continent, Gondwanaland, for the previous 60 million years.[10] Even today, New Zealand's native mammals consist only of marine species that swam or bats that flew there. Early settlers, however, rectified this perceived deficiency by importing a vast array of European species: red deer, Himalayan thar, chamois, fallow deer, samba deer, sika deer, rusa deer, stoat, weasel, ferret, rabbit, hare, and domestic cat. From North America came white-tailed deer. From Australia came a species of wallaby and the brush-tailed possum.

Ecological disaster! With the landscape full of food for these immigrant herbivores, their populations exploded. The forests could not withstand their intense browsing pressure, as no balance had evolved between the vegetation and herbivores that is common elsewhere. Moreover, without previous mammalian browsers to contend with, the native trees and shrubs had not developed spines or extensive chemical defences, as they had on other continents.[11] Whatever impact had been levied by the great flightless birds of New Zealand ended with the functional human-caused extinction of the giant moa in the 1400s. The result was denuded mountainsides, spring runoffs, and flooded lowlands, especially on the southern island.[12] While we lived there, we read newspaper accounts about the need to dredge this reservoir or that one because along with floodwaters came loads of silt, the product of mountain erosion. It was a lesson in the first law of ecology, that, directly or indirectly, "Everything is connected to everything else"![13]

For our study, we lined out the introduced species in a table and listed their characteristics, gleaned from an extensive litera-ture. We categorized productivity, or the reproductive rate, as high, medium, or low; the rate of spread after introduction as "fast" or "slow," with one mile per year being the separation point; and habitat selectivity by the number of habitats occupied.

The result? Limited correlations existed among the suites of characteristics typically believed to group together. For example, European rabbits had the highest productivity – up to fifty-six young per year – as well as a fast rate of dispersal, but the number of habitats occupied was only two. Red deer, in comparison, had the expected low productivity but exhibited a fast rate of dispersal after introduction, and a wide habitat choice.

Our findings suggest that the degree of correlation of traits in species appearing either early or late in succession has been over-interpreted. However, remembering that each species exhibits an *average* fitness among characteristics (as illustrated with moose in Chapter 4), variation in the packages of traits is not surprising. No species can maximize itself for every environmental need. Still, early-succession forests tend to select for individual pioneer traits to the extent possible, predominating in species like mice and rabbits, whereas more stable forests tend to select for conservative traits, as in woodland caribou and pine martens. The ecological literature is full of such examples.[14] And thus, healing comes to a disturbed ecosystem.

Community (or ecosystem) properties do emerge at different stages during succession, and they are important. They exert selection pressure that influences the number and types of species present, which makes the ecosystem resilient. This conclusion, though, raises a significant question. Throughout succession, do the communities exert selection pressure on their individual member species, or conversely, do communities exert selection pressure on themselves as composite entities? The difference is subtle yet profound.

By analogy, consider two hockey teams, the Edmonton Oilers of the 1980s and the Detroit Red Wings of the late 1990s and early 2000s, each with multiple Stanley Cup wins. The Oilers were a free-flying hockey team. The Red Wings, on the other hand, emphasized defence, requiring their forwards to circle back to mid ice any time the opposition got the puck in its own end and to employ "the trap" – a phalanx of players that the opposing team had to penetrate. A new player recruited to either team either fit in with the style of play or was rejected. This situation is analogous to the community exerting pressure on the individual.

However, there is another way to view the two hockey teams. Here the team gets together and concludes something to the effect, "Hey, we are an offensive-minded team and all we do is lose in a league of defensive teams, so let's change to be defensive, too." The team, in other words, exerts a selection pressure on itself *as a team.*

Here is a more controversial way to look at community-level selection. The target of natural selection is not the individual, as is generally accepted, but the community. Until recently, the only substantive debate has been whether genes, rather than individuals, are involved in selection.[15] Little attention has been given to the possibility of the target being higher levels of the biological hierarchy. Considering the detailed scrutiny that natural selection has undergone, and the ways it has been kneaded and massaged since first formulated, it seems strange that such a fundamental issue should remain controversial.

Granted, some adjustments in thinking are necessary. It is difficult to envision the mechanism by which natural selection can work on a target at a higher level than the individual. With individuals, you can think of selection affecting birth and death rates, thus allowing the best fit to leave the most offspring. Ecosystems have neither precise generations nor boundaries. They do, however, qualify as biological entities with their own integrity and characteristics, and so should be candidates for natural selection. Possibly, instead of the test of leaving the most offspring, more appropriate for ecosystems is the length of time they *persist*.[16] Persistence – continued existence for the longest time possible – of the best-fit ecosystem.

But whether the distinguishing characterisic is most offspring or persistence, not everyone agrees with community-level selection. In fact, most biologists still bristle at the concept. Not so the American evolutionary biologist David Sloan Wilson, who first published evidence for what he calls "multi-level selection" in 1997. Commenting on how long it has taken for this concept to be accepted, he wrote, "In the 1960s, a consensus emerged that natural selection almost never operates above the level of the individual."[17] He also noted, "Most [biologists] regard the concept of community 'superorganisms' as unlikely, bordering on the heretical."[18]

A few early ecologists sniffed around the idea of community-level selection when describing community-level traits, but none couched it that frankly. Now, however, an outpouring of journal articles has tackled the issue head on. The contention, as expressed by Sloan Wilson and co-author William Swenson, is that, "When selection acts at the level of whole communities, the community becomes analogous to an organism and the constituents of the community become analogous to genes within the organism."[19]

One article that has attracted considerable attention reports on artificial community selection done in the laboratory. The experiment was designed to see if different levels of plant productivity could be achieved after sixteen generations of selecting soil communities that supported the most growth from the least. The soil communities were true communities consisting of many interacting microbial and invertebrate species. Seeds were planted in containers filled with soil, and after thirty-five days the researchers separated those in which the plants did best from the ones that did worst, then used the respective soils to inoculate another set of containers to form the next generation.

The result was a four-fold difference in plant productivity.[20] Here was convincing evidence for the operation, in tandem, of soil species in communities. Before discounting this research because it was performed in a laboratory, remember that an important part of Darwin's rationale for natural selection was based on evidence gained from artificial selection by plant and animal breeders.

In 1997, one entire supplement of the notable journal *American Naturalist* was devoted to multi-level selection. In it, conclusions were drawn such as, "Multilevel selection theory has come of age and has revealed a fascinating world in which natural selection operates at all levels of the biological hierarchy."[21] Another conclusion was, "It is reasonable to expect higher-level units [communities] to evolve into adaptive units with respect to specific traits."[22]

Then, in 2002, a paper appeared with a provocative title that brought the topic of community selection right back to the issue of resilience: "Ecosystem adaptation: do ecosystems maximize resilience?"[23] The conclusion is, "Ecosystems evolve to the state most resilient to perturbation."

Shining through the uncertainty about how natural selection operates during succession are some simple facts. Even without human damage, almost all ecosystems are perpetually in a state of pending catastrophe. Soon they will be calling on whatever inherent resilience they may have. For some, like the boreal forest, naturally occurring fire happens normally every fifty to two hundred years.[24] Dry mountain forests may be fire-free for only twenty-five years. Prairies and marshes, too, burn on the scale of decades. Historically, what wasn't lit up by lightning was set afire by Paleoindians, who, all across North America, learned that young ecosystems meant a greater abundance of wildlife.

Not only fire, but insects, wind, floods, and avalanches play havoc so often that "catastrophe" loses its negative connotations and means only "sudden change," an integral part of ecosystems. The older a forest, the more vulnerable it becomes, with roots weakened by decay and broken branches littering the forest floor until collision with catastrophe is inevitable. Summing up an exhaustive review of North America's *Ancient Forests* as experienced by first Europeans, author Thomas Bonnicksen wrote that, "Pioneer species dominated the majority of America's ancient forests [forests that existed long ago], not older species, and the forests were generally open and sunny."[25] We only cling to a myth that, before European influence, all North American forests were dark and primeval.

With disturbance such an integral part of life, it is no surprise that ecosystems have adapted, that succession has written the rulebook, and resilience is the result. How could it be otherwise? Periodic death and rebirth of ecosystems is the way of things, generation after generation, just like the lineage of any individual species. Natural selection dictates its inevitability.

Consequently, deer are in the hills and grouse in the forests and a rich bounty of wildlife perpetuates itself across the land. Periodically, ecosystems are restored to highest productivity, firing on all cylinders as chloroplasts gather maximum sunlight. Nutrient cycles speed up. Forests throb with vitality; wetlands pulse with life.

Then there are humans. Nature's resilience has forgiven many of our land-use sins, too. But we seem bound to design experiments to see how far ecosystems can be degraded before their resilience fails and they slip into different "zones of attraction," where recovery back to what came before is impossible. There are limits. Have the marine ecosystems off the east coast of Canada and in Alaska's Bering Sea been irreversibly damaged by over-fishing of keystone predators such as Atlantic cod and Pacific pollock? Once keystone species such as these no longer play their ecological roles, cascading changes in other species completely reshuffle the biotic deck. In the journal *Ecology*, in 2006, is the obvious statement: "In the face of increasing threats to global bio-diversity, ecologists confront the challenge of understanding and predicting how such changes will impact community and ecosys-tem properties [and] the ability of populations and communities to *persist* through time."[26]

In the biggest human experiment of all, climate warming threat-ens to sweep cities off the flood plains like so much flotsam, kill many people, wreck the economy, and maybe set civilization back hundreds of years. Certainly such sudden change will restructure life on Earth. Resilience at this scale rests not only on succession but on natural selection, too – new species arising to fill vacant niches, just as they do after natural mega-catastrophes.

In the academy of life, always, some species are losers, others are winners. Where will humans fit in? Similarly, some ecosystems will fare better than others. But ecosystems are resilient. Maybe they will be rejigged, but they will bounce back, with or without humans.

CHAPTER FOURTEEN

THE PTARMIGAN'S DILEMMA

An overwhelming marvel is how the game of life has been played so vig-orously for close to 4 billion years without the ball ever being kicked out of bounds. Some finger-biting goal-line skirmishes have occurred. Knowing how species have kept the ball in play provides perspective on the length of time left before the final whistle.

ALL SPECIES ON EARTH are sandwiched in a "habitable zone" – a region of space where conditions for life are favourable. Sometimes it is called the "Goldilocks zone," being neither "too hot, nor too cold, but just right." Playing key roles in delimiting this zone are two conditions: adequate energy and tolerable physical environments. From an astronomer's perspective, their margins are set so close that the odds of life occurring anywhere seem slim.

On the unforgiving northern tundra lives a bird, the ptarmigan, capable of withstanding some of the coldest conditions our planet can dish out. Surviving on the edge of life's habitable zone, this remarkable bird is one of about a dozen species that experience full Arctic winters. Helping it to do so is a set of finely tuned adaptations, including feathered moccasins, downy underwear, and an overcoat of thick, insulating body fat.

So effective are the ptarmigan's adaptations that winter is not its period of maximum stress. Instead, the critical time comes when the snows are melting and the Lapland longspurs are sprinkling the tundra with song. That is the time of the ptarmigan's dilemma.

For three summers we pursued this fast-flying, chicken-sized bird up and down – mainly up, in our memories – three-hundred-metre (thousand-foot) slopes spread across a maze of tundra hills in central Alaska. Armed with nets on long metal poles, our objective was to run the ptarmigan down before they flew for a first, second, or third damnable time. When we were lucky, one of our English pointers would lock on a bird frozen motionless only centimetres from its nose. We would race there, stumbling awkwardly over the uneven ground, and clap our net over the bird before it could lead us on one of those breathtaking chases. We colour-banded its legs for later re-identification, and on some birds attached a tiny radio transmitter with a whip antenna that curved over its back. The harness and radio were ultra-light, and the birds showed no ill effect.

Ptarmigan, members of the grouse family, come in three species: rock, typical of grass-sedge tundra; willow, of shrub tundra near the treeline or in well-watered valleys; and white-tailed, of more southern mountaintops. All three tough it out on their breeding grounds all year. Our subject was the rock ptarmigan.

After strenuous ptarmigan-chasing days, we would retreat each evening to a log cabin in a stand of stunted spruces down at treeline. A simple wooden cross, which we found rotting on the ground and propped up outside the window, served as testimonial to the nameless cabin builder. At the cabin, we entered data, built telemetry harnesses, and repaired equipment. And in the late-night twilight, the haunting songs of gray-cheeked thrushes lulled us to sleep.

Spring is often cold in central Alaska, and even while the winter snows are melting, the wind can whip up blizzards, tundra ponds can refreeze, and the vegetation can linger in largely pre-green-up brown. This is when the all-white male ptarmigan parcel out the tundra into territories and display, to both rival males and prospective mates, with spectacular high nuptial flights and guttural calls. Bright combs flush red above their eyes.

The hens, with more to gain from protective coloration, moult from winter white to tundra-matching brown. They lay and incubate eggs, their chicks hatch, and still the temperatures commonly dip below freezing. The males leave the females to shoulder the task of rearing the next generation alone, and it is then that the females face the dilemma.

Ptarmigan chicks are "precocious," meaning they do not sit around in the nest waiting to be fed. Clutches can be as large as ten or eleven eggs, characteristic of grouse in general – too many chicks for a hen to feed. The chicks must find their own food or starve. However, there is a problem. Not yet able to regulate their body temperature, they must be brooded by the hen, who transfers heat to them from a bare "brood patch" on her belly. In a typical foraging session the chicks race out from under the hen, scatter in all directions, and peck vigorously at any suitable food they find. Then the hen gives a series of treble, *seepy* notes which summons the chicks back to be brooded.

But sometimes the clouds come down to race across the tundra. The winds pick up and drive a horizontal blizzard. The chicks have no place to hide, and besides, they must eat. The hen is faced with a dilemma. She must let the chicks forage long enough to meet their high energy demands but not long enough to exceed their tolerance for low temperature. Misjudgment will mean death for her chicks and biological failure for her.

The hen faces this dilemma with calm, genetically programmed passivity. More than one-third of the population of breeding females typically are yearlings, with no experience to draw on. Hens possess built-in environmental response mechanisms, but they are set, logically, by natural selection to average spring weather conditions. Unless these mechanisms allow for some behavioural adjustments in years of below-average conditions, chick production will suffer. And suffer it does; evidence exists of wide variability in the number of chicks produced year to year, not only in the Alaskan population but in Greenland, Iceland, and northern Russia. Periodic or partial failures in chick production may place long-term limits on the size that ptarmigan populations can achieve.

Our hypothesis was that, in severe weather, a ptarmigan hen places a priority on brooding her chicks frequently enough to keep them from dying from exposure, but the consequence is that they do not obtain sufficient energy. To test this premise, we had some technology at our disposal: the radios, with their whip antennas, worn by various hens; a bomb calorimeter at the University of Alaska's Institute of Arctic Biology; and the field biologist's ultimate weapon – a pair of binoculars. By sitting on a tundra knoll like frozen mummies we could monitor the radio signals to determine how long the hens allowed their chicks to forage under different weather conditions. When the chicks foraged, the hen did too, causing the whip antenna to jerk with treble distortions each time she pecked. When the chicks were being brooded, her respiration induced a gentle, rhythmic electronic wow.

With careful observation, sometimes carried out under the glow of the midnight sun, we were able to determine the pecking rate of chicks, which averaged one peck per 1.7 seconds during their brief periods of freedom. We also observed what they were eating and collected samples. Then we drove to Fairbanks to use the bomb calorimeter. Each type of food was dried, ground into a powder, weighed, and placed in a thick-walled flask with a lid that we wrenched tightly closed. The flask was filled with pure oxygen and placed in a water bath of known temperature. The apparatus was securely bolted down; failure to do that would have resulted in the lid being shot through the ceiling into the room above.

At the touch of a button, a spark ignited the contents. The temperature change of the water bath represented the energy in the food items, and with a series of equations could be converted to calories. Flies ranked highest, at 5.7 kilocalories per gram dry weight, an accomplishment for which we rarely give them credit. Large crane flies were the chicks' fly of choice. Next were the bulbils of a seemingly insignificant little pink-flowered plant called alpine bistort that the chicks obviously searched out, at 4.8 kilocalories, followed by the flowers and overwintering berries of mountain cranberry and blueberry. These items, along with a smattering of moss capsules and butterfly larvae, provided concentrated sources of energy for the chicks.[1]

Knowing from our observations that in life-threatening weather the chicks were allowed to forage a total of a mere 96 minutes per 24 hours, and that on average the chicks filled their crop in 82 seconds, and that a full crop averaged 0.47 kilocalories, we calculated that they could obtain 33 kilocalories per 24 hours. To compare that with how much they needed, we cheated. We turned to the poultry literature, because, other than one partially useful study of blue grouse, nobody had figured out the relevant energetics for wild grouse. Fortunately, lots of chicken researchers had made those calculations, and because body weights and metabolic rates were similar, our extrapolations were justified.

The answer? Even in the worst weather, the ptarmigan hens provided the chicks with sufficient time, just barely, to meet their energy needs. The chicks did not starve.

What about the other half of the dilemma, exposure to cold? Some chicks from a captive population that we raised from eggs helped us with that. Fitted with a temperature probe carefully pushed down their throats into their crops, they were placed in a cold chamber. The temperature was lowered to the point that they began to show their weak ability to counteract it by shivering. It was likely an unpleasant experience for them, but we took the chicks out before it was too late and returned them, agitated but unharmed, to their pens.

The data we amassed gave us some indication of what might represent critical temperature conditions. We then turned to Alaska State weather records to identify years when such conditions existed, looking for any relationship with chick survival. A decade of data on chick survival had been collected by Alaska State biologist Bob Weeden. He was somewhat of an Alaskan legend, often seen as a distant figure tramping the tundra hills around Eagle Summit, catching and banding ptarmigan. He drove his government-issue orange truck at motion-sickness speed on the curvy Steese Highway, causing locals who lived in the hills to time their travels to non-Weeden times. Many people thought his work was irrelevant – who cares how ptarmigan manage to live in the Arctic? Then oil was discovered on the north slope of Alaska at Prudhoe Bay, and because of provisions of the U.S. National Environmental Policy Act, an instant demand sprang up for an ecologist who understood arctic environments. One day, Weeden, who was also president of the Alaska Conservation Society, was called into his superior's office and informed that he could not continue as president and be a state employee at the same time. He quit his job. The Sierra Club snapped him up, and over the next few years he played an important role in delineating the environmental impacts of the proposed pipeline.

We ran Weeden's weather data on a computer, looking for a correlation with the data on ptarmigan chick survival. There was none. We tested various wind-chill indices from the climate literature, and a few we invented, with the same negative results. Years with the most severe weather did not result in reduced chick survival. Reluctantly, because of all the work we had put in, we concluded that ptarmigan had evolved the necessary tolerance and behavioural adjustments for the temperature conditions they encountered. Their periodic years of low chick survival had some other cause, later determined to be quality of the stock (Chapter 9).

The ptarmigan's dilemma is life's dilemma: to stay within the habitable zone. Through natural selection, all species have made, and will continue to make, modest but vital adaptations to do so. However, there are limits, and we inhabitants of Earth are fated to ride on our lucky planet like a jockey on a runaway horse, hoping the day is far away when Earth runs off the track, as it inevitably must.

Consider just one of these limit-setting conditions: energy. Its sources for life on Earth are scanty. One source is chemical synthesis that involves the oxidation of sulphides. This ancient source still energizes some specialized groups of bacteria, blue-green algae (cyanobacteria), and archaeans living either in deep-sea vents, hot pools, or cool anaerobic ponds. A similar energy source is the fermentation of organic molecules that have rained and continue to rain down on the oceans from outer space. Fermentation appears to have energized the very earliest unicellular organisms floating on or near the ocean surface.[2] As well, in a process similar to fermentation, some methane-producing unicellular organisms called methanogens are energized by combining carbon dioxide and hydrogen.[3]

Otherwise, it is only the radiation from the conversion of hydrogen to helium in our sun that energizes the habitable zone. Only one narrow portion of the resultant electromagnetic spectrum, visible light, is relevant for life-producing photosynthesis. And amazingly, only a few molecules on Earth, or more correctly a few closely related groups of molecules – chlorophylls, and to a lesser extent carotenoids – are capable of harvesting light energy and transforming it into chemical energy ready for use by plants.[4] (Another molecule has been discovered more recently whose significance is still being scrutinized. It is discussed in Chapter 15.)

Was the snagging of energy by planet Earth just a fortunate happenstance? It all seems impossibly lucky, but that may be a delusion. George Wald, the 1967 Nobel Prize – winning Harvard biochemist, would argue that the structure of atoms and the laws of physics preordained its occurrence.[5] Wald, who died in 1997 at the age of ninety-one, is a towering figure in the history of science, not only for his work on the pigment rhodopsin that allows for vision, but also for his own vision that extended from the world of molecules to the workings of the universe. He began one of his standard lecture series with the topic "The Origin of Life," and the follow-up series with "The Origin of Death." He is quoted as saying, "I have lived my life among molecules. They are good company." But he thought beyond molecules to humans and how we are constructed, with the famous quote that sums up how evolution works: "We are the products of editing, rather than authorship."

By editing, Wald meant the slow, winnowing process of natural selection acting over long periods of time. Even complex molecules like chlorophyll are constructed that way, by trial and error, through the fitting together of atoms in this way and that. If the molecule fails to capture light energy, it obviously cannot become a basis for life; if it works, it can. If the substitution of an atom or a subsection of the molecule works better, then it replaces the former one. Gradually, efficiency improves.

According to Wald, there is an inevitability about this process, partly because it involves a gradual assembly, but also because other molecules did not have the right prerequisite characteristics. The example of the properties of chlorophyll are instructive. Wald concluded, "There must come a stage in the development of life *wherever in the universe it persists* [our emphasis] when it must go over from feeding upon the accumulation of organic matter to synthesizing its own with the light from stars. When that time comes it seems to me likely that the same factors that governed the exclusive choice of the chlorophylls for photosynthesis on the Earth might prove equally compelling elsewhere."

But while we can praise natural selection for building chlorophyll, and be thankful for it, the very mechanism of photosynthesis is profoundly inefficient. Biologist Simon Conway Morris has observed, "The process of photosynthesis is far from perfect. There are many chemical steps, and if an engineer had been in charge of the design process, the accountants and administrators would by now have been asking awkward questions."[6]

Of the light energy falling on Earth, only about 10 percent is harvested by plants. The rest bounces off as albedo (the fraction of solar energy reflected back into space), or is absorbed as heat. Of the harvested energy, after metabolic costs, only about 1 to 3 percent remains to be incorporated into plant tissue. If some more efficient biochemical procedure than photosynthesis had evolved, with light-energy capture and subsequent conversions in plants up around, say, 20 or 30 percent, then how much more luxuriant life would be! But life on Earth is due to trial-and-error processes, so we should be grateful for what we have.

However, chlorophyll cannot capture energy without two everyday molecules that react with it in photosynthesis, and here the habitable zone is exacting. We take these two molecules – water and carbon dioxide – for granted, yet they will not always be present in amounts and forms that will allow the continuance of life. Planetary scientists looking elsewhere for life in the universe are able to narrow their search significantly to the region around ours or any other sun where these two molecules could allow photosynthesis to occur.[7]

Water is not only a solvent but also such a participant in the structures of life, even beyond its use in photosynthesis, that some biochemists consider it a "biomolecule."[8] While water is reasonably common in the universe, it is mostly in the form of ice. For life to evolve, it must be a fluid, which thus requires an appropriate range of temperatures.

For any planet, including ours, to be and stay even partly within that fluid range is a long shot. So close are we to the temperature boundaries of the habitable zone that in Earth's history are recurrent periods when we actually fell onto its very margins. About 2.3 billion years ago, the Earth was plunged into its first deep ice age, becoming, for the next 35 million years, a giant snowball.[9] Glaciers extended from the poles to the tropics, covering land and water alike. It happened again at least four times between 750 million and 580 million years ago,[10] and each time the ice nearly shut down the planet's entire productivity. Deep seas became anoxic. At the time of the first global glaciation, life consisted only of marine-dwelling bacteria, archaeans, and cyanobacteria. When later global glaciations happened, multicellular marine life was in its infancy. In all cases, life may have survived only at hot springs on the sea floor or in equatorial regions where the ice was thin enough for light to penetrate and allow limited photosynthesis to occur.

Snowball Earth is still being investigated, as is a less severe variant called "Slushball Earth," along with speculative conditions that plunged the planet into these deep freezes and managed to pull it out again. Less severe glaciations occurred about 440 million years ago on the part of Gondwanaland that is now northern Africa, at that time situated near the South Pole. It happened again 330 million years ago, more broadly across much of Gondwanaland, and much later, 40 million years ago, in the Antarctic.[11] In contrast, land masses in the northern hemisphere were never glaciated again after the snowball events, until the recent fluctuations over the past 5 million years. None of these recent glacial events threatened to extinguish life in the way the five earlier ones did.

Over the past 2.5 million years, three climate-altering cycles have interacted to cause periodic mini ice ages – seventeen of them, according to one source:[12] the 105,000-year Milankovitch Cycle, caused by variation in Earth's orbit around the sun from near circular to elliptical; the 40,000-year Obliquity Cycle, caused by variation in the tilt of Earth's axis; and a 23,000-year Precession Cycle, caused by wobble in Earth's orbit.[13] Because these cycles are built on enduring orbital features, they persist. While they do not threaten life on Earth, they can be expected to play havoc with the location of the habitable zone by interacting with solar luminosity, greenhouse gases, ocean currents, and photosynthetic rates to swing climates repeatedly from ice ages to hothouses. In the future, on whichever continent you stand, in all but equatorial regions, the climate will be both dramatically warmer and colder for periods of thousands of years – icecap to desert.

Unlike the icebox conditions that periodically threatened life, warm conditions have never done so, except very early in the planet's existence. While we would not want to blunt present-day concerns over global warming, global temperatures have been considerably warmer than today during all but the glacial periods mentioned. Each time, the increased temperatures enhanced life. Warmest was the Cretaceous Period near the end of the dinosaurs' reign some 100 million years ago,[14] and it caused life to flourish. Flowering (deciduous) plants first appeared, and trees grew in both polar regions as they did for the following 50 million years.[15]

Not only has temperature of most of the planet stayed within the habitable zone over the ages, so has the other component of photosynthesis, carbon dioxide. Once, 15 to 10 million years ago, atmospheric carbon dioxide may have fallen perilously close to the critical level for photosynthesis, ten parts per million (ppm). That crisis was caused, in part, by the uplift of the Himalayas, which sopped up atmospheric carbon dioxide through the weathering of newly exposed rocks.[16] Otherwise, levels of atmospheric carbon dioxide, while on a roller coaster over the eons, have remained within habitable limits. Currently, the atmosphere contains 370 ppm, having risen from 270 ppm in the pre-industrial late 1700s. In comparison, the level was as high as 500 ppm about 450 million years ago, staying high for most of the 290-million-year Paleozoic Era, before dropping briefly to a low of 33 ppm about 280 million years ago.[17] Over the past half million years, levels have roughly cycled four times, hitting highs of about 300 ppm and lows of about 160 ppm.[18]

In summary, almost unbelievably for a lonely planet spinning madly in space, for almost 4 billion years the conditions for photosynthesis, energy capture, and life have stayed within the habitable zone. No frigid atmosphere, as on Mars. No runaway greenhouse effect, as on Venus – not yet, anyway.

Energy capture is one factor setting narrow limits on the habitable zone; another is the ability to deal with it after it is caught. Energy channels life into trophic levels: plants, herbivores, predators, predators on predators. Each level is only about 1 percent efficient in passing energy to the next level – a humble accomplishment. Consequently, terrestrial ecosystems usually have only a frugal three or four levels. Efficiencies are slightly better in aquatic or marine ecosystems so there may be another level or two. Best is a marine-terrestrial interface, which has as many as six levels.[19] Eventually, all energy is lost as heat.

Even though inefficient, the flow of energy into ecosystems manages to organize them in ways that magnify their hold on life. Energy relationships dictate the body sizes of animals, thus expanding the array of exploitable niches available in ecosystems. This expansion strengthens ecosystems and helps them persist. It seems as though ecosystems hang by their fingernails over a cliff of insufficient energy, but their fingernails are neatly fashioned to not be dislodged.

As illustration of body size expanding the number of exploitable niches, for more than a decade we studied and directed the work of graduate students on the implications of body size in wolves, coyotes, and foxes in the southwest Yukon. Each of these predators lived or spent time in our core intensive study area, the Duke Meadows, which consists of a series of wide gravel bars laid down by the turbulent Duke River and covered with a mixture of grasses, shrub poplars, willows, and spruces.

Our headquarters, when not away at some distant den or caribou plateau, was the Arctic Institute of North America's Icefield Ranges Research Station, a cluster of unimposing huts scattered in the willows beside an airstrip. To the west rose the rugged and rocky Kluane Ranges, to the east the rolling tundra Ruby Range, while to the north lay glacier-blue Kluane Lake. Amid this beautiful scenery we engaged in the menial task of collecting scats by the sackful for hair and bone identification.

We found that these three canid predators, each with a different body size, presided over different, albeit slightly overlapping food webs. However, all three predators periodically foraged in the nearby Burwash village dump, a perilous place for them to be. One day, graduate student Chris Wedeles idly dialled up the frequencies of missing foxes and coyotes from a knoll above the village, only to discover that their radio collars were down below. By walking through the village waving his antenna, which drew a crowd of curious children, he was able to identify the specific houses where the collars were. The animals had been shot and the collars kept as curiosities.

The dump was a fast-food outlet used occasionally by the canids. Their main dining hall was out in the wild, and together, the three different-sized canids preyed on most of the herbivorous mammal species living in the region, from mouse to moose. Snowshoe hares comprised the bulk of both fox and coyote summer diets, supplemented with arctic ground squirrels and voles.[20] However, the two predators caught snowshoe hares largely in different habitats – foxes in the forests, coyotes in the grassy or shrubby openings. Wolves ate primarily moose, caribou, and beaver.[21] It was a nice division of resources.

This array of body sizes exists in mammals because of a fundamental energy relationship. Small species have fast metabolic rates, much faster per unit of weight than those of large mammals. Small herbivorous mammals, therefore, must exploit high-energy sources such as seeds and berries. They live life in the fast, energetic lane, but for reasons of energy conservation stay in relatively small areas. For large species, the reverse is true. They can exploit lower-energy food sources like leaves and twigs because they have longer guts, and so, as a general rule, can extract more energy out of their food. The intermediate-sized species fall between, with energetic and territory-size characteristics at intermediate levels.[22]

With this array of animal sizes, all built as contrivances that capture various sources of energy, ecosystems can support a rich diversity of species. Maximum use is made of the limited energy that flows through the ecosystem. Animal species of various sizes are patterned within the ecosystem like boxes within boxes, small territories within larger territories, with energy directing the way the ecosystem is put together and making the best of its inefficiencies.

———

Living things do not just passively receive energy, of course. They adapt to use it most efficiently, and that ability also helps keep them in the habitable zone. Because energy capture is seasonal, whole ecosystems respond with an annual "productivity pulse," a spontaneous, sudden eruption, like the enthusiasm of hockey fans reacting to an overtime goal. Annual productivity varies among the world's terrestrial ecosystems, being highest in tropical rainforests. Temperate deciduous forests come in 40 percent lower, and boreal forests another 33 percent lower again.[23] Yet, the sudden productivity pulse, the period in the year when the rate is highest, is most pronounced where temperate deciduous and boreal forests meet – not in tropical forests, where energy capture goes on most of the year, or on the tundra, where the growing season is short. Most dramatic in North America, in our experience, is the hectic five weeks from late May until the end of June on the edge of the Canadian Shield in eastern Canada and parts of the northeastern United States. The winter lockdown is over. The forests seemingly explode with joy.

For several years during the productivity pulse in Ontario's Algonquin Park we measured the growth of the leaders of sapling aspens, maples, birches, and balsam firs, the primary food of moose. We did this in cut-over forests, shouldering through dense, regenerating saplings following pre-set, random compass bearings. This was normally hot, mind-numbing work. Our task was to determine how much annual new growth, and hence palatable and nutritious forage, the forest was producing for moose or deer, and if that availability of food correlated with their survival the following winter when conditions for them were harsh.

It was a research dead end. We discovered that twigs chewed off by a moose or deer are almost never the full growth from that year, just a quarter, or half, or three-quarters – rarely more. Because the animals do not exploit all there is to exploit, differences in the amount of growth in different years must be largely inconsequential. So, we never did analyze the thick file of data.

While doing this work, however, we discovered that virtually all the year's new growth for these tree species is completed by the end of June. Trees continue to capture energy – they must, because green leaves in sunlight are programmed to do so – but the energy follows different metabolic pathways into maintenance, or more significantly into storage for winter. For deer and moose, and indirectly wolves, most of the energy input that keeps them in the forest occurs during this brief, critical productivity pulse.

In the Arctic, the productivity pulse is so brief that in some years it does not even happen. That was the case in 1999 on Banks Island in the western Arctic. That summer we canoed the Thompson River in recently created Aulavik National Park. The north-flowing Thompson cuts an increasingly wide swath across the treeless island for some 250 kilometres (155 miles), ending in an ice-choked arm of the Arctic sea.

The airplane dropped us off as far upriver as the previous winter's ice would allow, then we paddled our way through rolling tundra dotted with muskoxen. We expected to see an abundance of wildflowers, as typifies the Arctic in summer, blooming in synchrony because of the season's shortness. Instead, much of the time we battled strong headwinds that built standing waves on the river and suffered the occasional whiteout from snow. The hills remained brown, and we stayed cocooned in our down parkas. Except on well-sheltered south-facing slopes, the tundra never did green up that summer. The muskoxen ate the dead stems of grasses and sedges from the previous year's growth. New energy input was near zero. Life in the far north lies dangerously close to the outside limit of the habitable zone.

As with energy, living things do not just passively receive whatever temperature conditions are thrown at them. While temperature adaptations are largely unique to individual species, genera, or families, some are sufficiently widespread to be categorized as "rules." Really, these rules are more like tendencies. Biology at the level of species and ecosystems, with its added complexities, must be innovative and flexible. Among warm-blooded animals, individuals within a species or closely related species tend to be larger in northern than in southern parts of their ranges, an observation made back in 1847 by German biologist Carl Bergmann. The reason relates to both heat production – larger animals have more cells that are metabolizing and producing heat as a by-product – and heat conservation – larger animals have less surface area per unit of volume and so lose less heat. Examples are many: white-tailed deer in Nicaragua are about one-third the size they are in the northeastern United States; each of the four subspecies of least weasel in North America is progressively larger as you go north; cougars, with their extremely wide range, are much larger in the Yukon than in Argentina; the largest subspecies of moose lives in Alaska, the smallest in Montana.

In 1877, American naturalist Joel Allen did Bergmann one better and formulated Allen's Rule. It states that the extremities of warm-blooded animals are shorter and smaller in northern parts of their ranges. This rule is most obvious when comparing closely related species like arctic hares, which have short ears, with desert jackrabbits' long, mule-like ears. The same size differences hold true for the ears of arctic wolves versus Mexican wolves, and arctic foxes versus red foxes. In these extremes, not only do northern species keep warm by conserving heat, but southern species keep cool by radiation.

These "rules," however, are not inviolable. The smallest, not the largest, subspecies of caribou is found on the high Arctic islands. Caribou beat the rule by having an exceedingly well insulated body with a coat of hollow, white hairs lacking central pigment. Red foxes have larger, not the expected smaller feet in the north, particularly where snow is greatest,[24] but gain the advantage of staying up better on tundra snows. Coyote foot sizes also are larger, not smaller, in the north, again violating Allen's Rule, but show no correlation with snow, which leaves them with no obvious northern adaptation. This situation may reflect their short history in the north. Only since European settlement of North America's central plains, the original home of coyotes, have they expanded their range, now all the way north to Alaska. Why no appropriate foot adaptation has been selected in the intervening years among the inevitable natural variation in foot size is unclear. Such adaptations can happen rapidly, as is the case with house sparrows, new to North America in the mid 1800s, which now exhibit both Bergmann's and Allen's rules – larger bodies with shorter extremities in the northern part of their range.[25]

In other important ways, species adapt to stay within the habitable zone. They follow climate change up and down the face of continents by shifting their ranges and opening new franchises in suitable localities. Mobility has always been an important antidote to extinction. In 2001, while backpacking in the western Arctic along the spectacular, canyon-enclosed Hornaday River with our daughter Michelle, a national park warden there, we suddenly flushed an ordinary, everyday robin. The Hornaday flows into the Arctic Ocean, and as we scrambled through waist-high willows only a few kilometres from the river's mouth, the robin flew a short distance, landed in a willow, and rolled out its familiar songs, just as robins do on lawns in downtowns anywhere. Robins are shifting their range to adjust to climate warming. This was one of the first recorded on the western Arctic coast.

Ghosts of species that shifted their ranges as far back as the glacial epoch ten thousand years ago still haunt the hills. Mountain sheep, for example, show the influence of isolation during the Pleistocene. The white Dall's sheep of Alaska and the Yukon took refuge in ice-free lands between Siberia and Alaska known as Beringia, while stone sheep, Rocky Mountain bighorns, and California bighorns survived on various mountain peaks or plateaus that protruded above the ice. Desert bighorns, in contrast, were pushed south of the ice. Similarly, caribou were separated into subspecies: Peary's, which lived in an ice-free area on Ellesmere and Axel Heilberg islands; grantii, which lived in Beringia; and the subspecies of woodland or mountain caribou that lived south of the ice. Even birds, despite their mobility, have twin eastern and western species counterparts, each pair obviously originating from the same stock: eastern and western bluebird, myrtle and Audubon's warbler, scarlet and western tanager, northern and Bullock's oriole. Genetics research has indicated that these pairings, once thought to reflect a dry habitat barrier in Pleistocene times running through central North America south of the continental ice sheet, are actually ghosts from an even earlier separation of some sort that happened up to 5 million years ago.[26]

The habitable zones of species today are changing along with the climate, and that provides rich fodder for research. We are already witnessing how temperature reshuffles the biotic deck – great stuff for science, if nothing else. Inevitably, life will be rearranged with winners, which can migrate or adapt, and losers, which cannot. But eventually and inevitably, all life will run its course, and the time will come when our planet is no longer habitable.

This grand departure of life will not happen because the temperature becomes too hot or too cold for liquid water – seemingly the most likely cause – but from a shortage of the other essential component for photosynthesis: carbon dioxide. Little known is a terrible irony about carbon dioxide. While today we concern ourselves with its human-caused atmospheric increase, someday we will run out. Before jettisoning a concern over greenhouse gases, however, realize that it will not happen for some time, between 500 million and 1 billion years from now.[27] Then, atmospheric carbon dioxide is predicted to fall below ten parts per million, the critical level for photosynthesis. Then, the terrestrial biosphere as we know it will collapse for a certain and final time. Deep-sea vent creatures may be all that are left. For the remaining 80 or 90 percent of its projected 5-billion-year lifespan, Earth will spin here without terrestrial life. Soil will wash away. The planet will become a rock and seascape once again.

We will run out of atmospheric carbon dioxide because its overall trajectory, despite fluctuations, is downward. Nothing will bail us out, not even the carbon dioxide produced by our belching automobiles and factories, which will turn out to have caused only a pitiful blip in the long-term scheme of things. Radioactivity in the Earth's core is continuously declining, and that is the source of the volcanic activity that is the primary source of atmospheric carbon dioxide.

If researchers turn out to be wrong about this rather well-supported prediction, we should not feel overjoyed. Rising temperature is waiting in the wings, and only the time scale need be adjusted. The sun's luminosity, and thus its temperature, has increased 25 percent since life began,[28] and will continue to increase. All stars become progressively hotter until they explode as red giants, and ours will reach that stage in approximately 5 billion more years. But well before then, between an estimated 1 billion and 1.2 billion years from now, heat from the sun will terminate all life on Earth.[29] Geophysicist James Lovelock predicts that only 100 million years from now the sun's heat will be too much for the Earth to regulate at its current state. The Earth will be forced to move to a new hot state possibly inhabited by a different biosphere, one without the plants and animals or even many forms of bacteria we know now.[30]

Long before the cataclysmic events described above, we intend to go back to Alaska to see how the rock ptarmigan are doing. Have warmer temperatures solved their dilemma by allowing plenty of time for the chicks to forage? Will they nest earlier because they are temperature- rather than daylight-programmed? And if so, will the polygonum flowers be blooming, and will the crane flies have emerged earlier, too? Maybe we will have to travel farther north to find rock ptarmigan, because shrub tundra and forest will likely have converted the hills into habitat for willow ptarmigan or even spruce grouse. How far north? How long until ptarmigan and tundra get pushed off the top of North America?

We published our findings about the ptarmigan's dilemma in the journal *Arctic*.[31] An insignificant step in the staircase of science, it did not win us a Nobel Prize. But the rolling tundra and the guttural calls of ptarmigan keep replaying in our memories, and for all those breathless chases, that is reward enough.

CHAPTER FIFTEEN

LET THERE BE LIFE

"And so we come to perceive life as a force as tangible as any of the physical realities of the sea, a force strong and purposeful, as incapable of being crushed or diverted from its ends as the rising tide."
    – Rachel Carson, The Edge of the Sea[1]

SOME PLACES IN THE WORLD are beautiful beyond words. They overpower your senses, render you speechless, strike at the very essence of your being. They gut out of life everything that is trivial. They humble you with their immensity, their dispassion.

Neko Harbour in the Antarctic is such a place. Tucked away in a remote corner of the world, until recent years it was unfamiliar with humans. Its geologic and biologic processes went on without acclaim. It had existed in various forms for over 40 million years, part of the beauty of the Earth where, on the geological time scale, humans are only an afterthought.

We arrived on a sparkling sunny summer day in late January.

Ian Sterling, a staff biologist aboard ship, had remarked that if there was one day on the whole trip that had to be sunny, it was when we sailed into Neko Harbour. He had been there before, and he was right.

What makes Neko Harbour beautiful is an overwhelming, rich bounty of wildlife on land, in the sea, in the air, animating a handsomely rugged, sprawling scene – 360 degrees of ice and rock piled in stunning disarray. Add to that a wide fjord of sea: blue, perfectly calm, studded with dazzling icebergs of all shapes and sizes. They had calved from the semicircle of steep-walled glaciers that poured out of the surrounding mountains, depositing their frozen cargo directly into the sea. We stood on the bow deck of the *Sergey Vavilov* as the ship's captain slowly piloted us into this palace of ice.

We were seeing the Antarctic for the first time, a place that had fired our imaginations since student days when, at the Institute of Arctic Biology in Fairbanks, we listened to the stories of biologists who were working on the southern continent. In common with them, we were interested in the cold-temperature adaptations in wildlife, and once we'd thought our careers would take us to the extreme south. Instead, we did not get there until 2008.

After a short ride in Zodiacs, we and our ninety-two shipmates stood on a rocky, ice-strewn cobble beach. Only then could we begin to appreciate the wildlife around us, most notably the penguins. Gentoos, sporting orange beaks and white earmuffs, in tight knots on snow-covered knolls that were painted pink with their droppings. Gentoos in dense colonies high up on ice cornices, with parades of birds waddling stiffly down to the water, passing others waddling back up, all intent on feeding their patiently waiting young. Gentoos exploring the water's edge, poking around the grounded ice, diving into the sea and cannoning out. Gentoos walking up to examine us at macro-lens distance. Gentoos throwing back their heads and uttering their penetrating jackass-like calls. Gentoos on a wild white-and-blue stage. Patrolling overhead glided giant skuas and petrels, as they do over all penguin colonies in the Southern Ocean. Big brown birds with beaks adapted to tearing flesh, they serve as both garbage collectors and executioners, forever on duty.

We stayed for a couple of hours, climbing up a snowfield for a better view. Our ship, anchored offshore, shrank to insignificance in the vastness of the scene, appropriate because anything human-made seemed so out of place. Appropriate, too, because the Antarctic continent is largely human-free, a 10 percent anchor to a world awash with mankind.

Later, we again loaded into the Zodiacs and cruised the fjord to become acquainted with some of its marine mammals: whales blowing, surfacing, and lobtailing, and seals hauled out on the ice floes – crabeater seals, Weddell's seals, and leopard seals – a top carnivore of the Southern Ocean. We slowly circled one ice floe while a huge leopard seal grinned at us, the corners of its mouth raised in an ironic, perpetual smile, the last thing seen by many a hapless penguin. Suddenly the arched back of a minke whale broke the surface, and seconds later we realized, from its dorsal fin slicing the water, that it was coming directly at us. At the last moment the whale dove, just deep enough to avoid bashing the bottom of the Zodiac. It swam directly beneath us, then, as if taking a curtain call, repeated the performance.

Other minke whales were spouting in various places among the bergs, the noise of their exhalations crystal clear in the crisp, still air. As we returned to the ship, one minke followed in the wake of our Zodiac until we were near the gangway.

———

All this rich abundance of wildlife is, in an ecological sense, the megafauna icing on an oceanic biological cake. The cake itself is obscure to terrestrial beings like us, a world apart. Yet we share the planet, and are energized by the same sun, partners in the dance of life with the same atoms that cycle endlessly on. Surely there was a message in this shared dichotomy.

We encountered that megafauna icing a dozen more times at other landings, on either the Antarctic Peninsula or the sub-antarctic islands of South Georgia or the nearby Falklands. We encountered it, too, in the Antarctic Ocean, with its circumpolar current, when we crossed two zones – the Subantarctic Front and the Polar Front – where cold, deep water flowing off the Antarctic continental shelf brings nutrients to the surface, especially iron, which is a limiting element in most ocean ecosystems.[2] These fronts are biologically rich places because there the nutrients mix with warm, south-flowing currents from the Atlantic, Indian, and Pacific oceans.

Along the Subantarctic Front came whales in abundance, especially humpbacks that surfaced, rolled, and sounded, with tail flukes raised high. Smaller minke whales were abundant, too, and, most exciting, a sperm whale, with its distinctive forward-slanting spout. Rich in symbolism, sperm whales – the whale in *Moby Dick* – are a testimony to species resilience, still swimming all the oceans of the world despite human slaughter of an estimated million animals.[3]

We crossed the Polar Front early on January 14, riding a restless sea that banged galley doors and made everyone stagger. Within an hour, the temperature had dropped from eight to four degrees Celsius, and although totally unrelated, the weak sun faded out. As if on cue, suddenly the air was full of seabirds careening in the wind. They tested our identification skills as we, too, careened around the plunging deck. Then, like a king among them, appeared the much-anticipated marvel of the oceans, the perfect seabird.

Evolution began a project to craft the perfect seabird back in the mists of time. It had to be adapted to the eternal winds and waves that make up 80 percent of the surface of the Earth. It had to be capable of gleaning a living from the barren-seeming ocean surface, be comfortable in gale-force winds, and be content to spend years out at sea far from the sight of land.

During the early construction phase, back near the end of the Mesozoic Era some 73 million years ago, evolution succeeded in driving a wedge between ancient shorebirds, with which it had been experimenting for some time, and a new, second group that it would eventually fashion for life at sea.[4] Not satisfied with the result, about 2 million years later evolution shook up this new group by splitting it again into what were destined to become the loons, penguins, and, in a whimsy of unpredictability, the storks, from the archetype of its perfect seabird. Through good luck or good adaptations, these early birds made it through the extinction catastrophe at the end of the Mesozoic. After that, evolution seems to have stalled, maybe busy whipping up early mammals to radiate into new species. However, it still played around with its seabird archetype, inventing various sizes and models.

Finally, about 15 million years ago, while working with a small near-shore prototype, evolution succeeded in fashioning the perfect seabird.[5] It gave it long, slender wings, longest of any bird at 3.5 metres, and wing-joints that lock so it can glide for hours, for days, with scarcely a wingbeat. The bird was, of course, an albatross. Then, about a million years ago, evolution created the king among them, the largest and best of the breed – the wandering albatross.

To make an albatross, evolution had to overcome formidable barriers. It had to work out a way for the creature to handle sea water, so it placed a tubenose on top of its beak that concentrates and helps excrete the salt. It needed to provide a navigation system for the bird as it skimmed over featureless waves, and an ability to find sparsely scattered patches of food – squid, fish, krill – so it gave the albatross large olfactory sacs, larger than any other bird's. With those sacs it was able to detect the macro chemical signatures of the seascape[6] and even the scent of krill and fish being fed on by other organisms.[7] Evolution had to include a capacity for making 10,000-kilometre-long foraging trips in some of the strongest winds on Earth, and accomplish that without draining the albatross of energy. It did that by endowing the bird with long wings and a gliding capability that raised its heartbeat over its resting rate by only 10 to 15 percent, instead of the few hundred percent of most other birds.[8]

With such marvellous adaptations, it was an exciting event when the first wandering albatross showed up skimming the waves behind our ship. So it was, too, for pioneer marine ornithologist Robert Cushman Murphy, who, on seeing his first albatross from a whaling brig in 1912, wrote: "I now belong to a higher cult of mortals, for I have seen the albatross! I have been watching the wonderful gliding of the grandest of birds during much of the day."[9] Those were the days when ornithologists collected everything they saw, however, and Murphy went on to write, "As soon as we have moderate weather, I think that we can hook some of the albatrosses and start my series of specimens."

Murphy was not alone in hooking albatrosses. Since then, many thousands have died, wanton by-product of longline fishing that involves dispensing lines of hundreds of hooks from the back of fishing boats. Up to 100 million hooks are set annually in the Southern Ocean.[10] In 2004 the remaining estimated population of 55,000 wandering albatrosses was classified as vulnerable by the International Union for Conservation of Nature. Since then, conservation efforts have met with some success thanks to new methods of sinking the hooks faster.

Our first wandering albatross was a magnificent, fully plumed adult with largely white wings edged in black. It stayed low, then glided steeply up off the stern deck and hung just above our heads. We could see its wings vibrating like violin strings in the wind. It slanted away, dropped back, yet minutes later was beside us again with never a wingbeat, just playing around in the forceful wind currents and the ship-generated luff-zone like a musician playing scales up and down a keyboard. Wandering albatross – bird of legend, hung around the neck of the ancient mariner as a sign of guilt for having killed it; bird that embodies the souls of dead sailors; bird of great beauty and grace; distillate of the invisible marine ecosystems.

Other distillates made their appearance, too. Much more common than wandering albatrosses were smaller, darker, black-browed albatrosses. We visited one of their colonies on the steep rocks of Carcass Island, an especially important member of the Falkland group because it is rat-free. A legacy of early explorers, sealers, and whalers was the introduction of Norway rats, intentional or otherwise, to islands in the Southern Ocean, as well as the South Pacific. The native wildlife on these islands had evolved no defences against such a land-based predator.

In our notebooks from this exciting day, as we crossed the Polar Front, are listed other sea specialists: light-mantled sooty albatrosses, which many ornithologists deem the most beautiful because of their dark heads and contrasting white eye-ring; southern giant petrels, including the startling all-white colour phase that cruised by the ship repeatedly; and dark-coloured, white-chinned petrels. Then there were the tiny, swallow-sized sea-specialists – Wilson's storm petrels and black-bellied storm petrels – adept at slipping between the wave crests and the gusts of wind, and in the habit of "dancing" on the water as they search for surface plankton, squid, and small fish. They can stay stationary even in a horizontal wind by the hydrodynamic drag of their feet through the water – their dance – which balances the aerodynamic drag of the wind.[11] As well, little nondescript birds called prions, almost indistinguishable because the various species are so alike, buzzed around, making it seem as though we were in a forest full of migrating songbirds, except here there were no trees. Only wind and wave.

All this abundant life was dependent on a world foreign to us, a realm we could not see. The light bouncing off the sea surface curtained what lay below. Sometimes the surface was dark and foreboding, sometimes sparkling, sometimes choked with ice, and sometimes whipped up by a wind that lashed the wave crests into spray. Always, the surface sequestered its unfathomable secrets. We could not see what the albatrosses and petrels saw as they dipped and dove over the waves. We only glimpsed the aerial world of whales and porpoises when they came to the surface for air. We lost track of the penguins and seals the moment they plunged off the icebergs. The food they sought was part of a web of life so different from the terrestrial one that we humans are familiar with that it could have existed on another planet.

Yet the biomass of life in the seas, the part we do not see, makes up at least half of the life on Earth,[12] a fact that has emerged only since the beginning of the twenty-first century. When the capability of marine exploration progressed from what could be caught with a dragnet to what could be caught with a fine filter and analyzed in genetics labs, we learned that bacteria and viruses exist in sea water in vast numbers.[13] An average millilitre is estimated to contain around 1 million bacteria and 10 million viruses,[14] as well as an unestimated number of cyanobacteria[15] and an array of Archaea that themselves could make up 50 percent of life in the open sea.[16] This represents a lot of life, because sea water is by far the largest habitat on Earth.

The number of individual microbial species in the sea, however, will never be known. Here is a colossal case where the concept of species breaks down. Viruses have the capability of carrying and distributing genes throughout the microbial community, maybe doing so on a massive scale.[17] Instead of searching for species, it seems more appropriate to search for genes, or the protein products of genes. Ocean sampling and genetic analysis conducted by Craig Venter and his team in 2005 found genes that may code for more than 6 million different proteins, about doubling the known number.

It makes sense that the oceans should be rich in life, richer than land, because life there had a head start of about 3 billion years. With greater space and time, it makes sense, too, that life in the oceans evolved a few unique twists. Common between ocean and land is photosynthesis, but the single-celled photosynthesizing diatoms that predominate in oceans, and reproduce asexually when nutrient and light conditions are right, can double their populations in just a few hours.[18] So can dinoflagellates, another group of photosynthesizers that often outnumber diatoms in warmer seas. Also, in an unexpected feat of adaptability, these and other organisms can capture energy not only in the sunlit upper layers of sea water but also in extreme places that have no equivalent on land, for example on the undersides of and within sea ice itself. There, in the microscopic brine channels that honeycomb sea ice, cold-temperature "extremophyles" prosper, grow, and reproduce. Speculation holds that this ice option kept life from blinking out during the periods of Snowball Earth.

But there is more. Life in the sea has worked out two other unique mechanisms of energy capture. Both have been discovered only recently, emphasizing how little we know. One involves the deep-sea vents that characterize the mid-oceanic rifts around the world and may have been where life originated on the planet.[19] The very existence of these vents was unknown until 1979, when first observed from the deep-sea submersible *Alvin*. The rich variety of life found there was astounding. Supporting this life are bacteria that capture energy without light, using sulphides instead of carbon dioxide in a chemical reaction analogous to photosynthesis. Both bacteria and archaeans accomplish this feat. Both are appropriately classed as extremophyles, living in extremely hot and anoxic water. They, too, may have supported life through the periods of Snowball Earth, tucked away at the bottom of the oceans. We now know that extremophyles can live in acid pools, nuclear reactor wastes, and rocks far beneath the Earth's crust.[20] They are a great backup to an uncertain future.

Moreover, a third route of energy capture was discovered so recently that its implications are only now being assessed. Possibly, it represents a major way that energy is captured on the planet besides photosynthesis. Proteorhodopsin (*pro-tea-o-row-dopsin*) is the name of a family of about fifty proteins that are related to the light-sensitive pigment rhodopsin, found in the retina of the vertebrate eye.[21] Laboratory studies have shown that light energy falling on proteorhodopsin creates an electric potential sufficient for metabolism and growth.[22] This process is analogous to the sun acting on a solar panel, except that protons rather than electrons do the moving, and the medium is a protein rather than a silicon chip. In 2000, geneticists were amazed to discover genes of bacteria, archaeans, and viruses that code for these proteins captured on filter paper samples of sea water.[23] Subsequent inventories have revealed that these genes are carried by about 13 percent of the vast billions upon billions of ocean bacteria.[24]

Investigations are underway already to evaluate the potential of this source of energy for human needs. But for understanding evolution, the discovery provides another way, besides fermentation and chemosynthesis, that bacteria and archaeans may have captured energy before chlorophyll-based photosynthesis became efficient enough to flood the seas and atmosphere with oxygen. While there is evidence that early bacteria were photosynthesizing 3.5 billion years ago,[25] it took another billion years for a significant volume of oxygen to accumulate in the atmosphere and oceans.[26] Respiration, the reverse of photosynthesis, sopped oxygen up at close to the same rate.[27]

However energized, all these simple marine organisms provide the basic first step for evolution to work on and fashion into webs of life, just as photosynthetic bacteria and plants are the first step in terrestrial environments. Parallels between marine and terrestrial webs are inevitable, because the principal force that drives them, natural selection, operates everywhere there is life. It cannot be otherwise.

Thus, the diatoms and other primary producers of the Southern Ocean are fed on by various single-celled herbivores such as foraminifera, whose abundance even in the brine channels can reach over 1,000 per litre of ice,[28] and small crustaceans, which are dominant in oceans worldwide. Next come the larger, omnivorous amphipods and crustaceans that both graze the phytoplankton and prey on the little copepods; then, king of them all, the six-centimetre-long Antarctic krill, *Euphasia superba*. Krill are estimated to exceed the human biomass of the planet three times over.[29] They are the keystone species in the Southern Ocean, because they feed so many predators either exclusively or to varying degrees – squid, many species of fish, and all marine mammals and birds, either exclusively or to varying degrees.

Even higher in the trophic pyramid, some species are predators on predators, such as the great toothed whales like sperm and killer. A leopard seal lying on an ice floe is truly a summit predator; it may have dined on penguins that have eaten fish that have eaten krill that have eaten small copepods that have grazed on diatoms.

Losses of energy to digestion and metabolism eventually put a cap on the number of trophic levels. Meanwhile, on the ocean-bottom wait the scavengers like flatworms, starfish, sea spiders, and sponges, on the lookout for organic detritus that filters down. Finally, always patiently biding their time because eventually all life comes to them, are the decomposers, the bacteria that turn organic matter back into its constituent nutrients. In time, ocean currents bring these nutrients back to the surface so that eager diatoms and other unicellular plants can take them for another fling on the dance floor of life. The same processes of energy flow and nutrient cycling happen everywhere.

Meanwhile, we rode the waves in our ship, aliens, out of place, dependent on what we brought from land to sustain us, foreigners in a world we hardly knew anything about. The scientist members of the crew increased interest and expectation among the passengers by giving lectures in a lower-deck, windowless room that pitched and heaved.

At mealtime we lined up to eat from our land-based food web, sometimes seeing icebergs go by the windows but nonetheless complacent in the normalcy of a dining room. We were not part of the food web around us. No krill on the menu. No leopard seal about to slide up behind us and pick one of us off. Even detritivores waiting on the ocean floor were not going to consume our nutrients. We were isolates. It was then that the irony of our tiny ship, a spot of terrestrial flotsam out there on the wide expanse of the world's dominant ecosystem, became most intense. We were briefly visiting this other world that makes up most of "our" world, but physically only skimming the surface, just as our knowledge merely skims the surface, too.

Other fundamental similarities, besides photosynthesis and food webs, operate in both marine and terrestrial ecosystems. The same process of succession happens in the sea as on land, and the principal driving force is the same, too: one community causing changes in the environment that invite in succeeding ones. Marine succession is more of an annual process than the multi-year phenomenon it is on land, and marine succession is influenced more by quickly changing physical conditions. Otherwise, succession unrolls in a similar way. For example, among the life forms that photosynthesize, diatoms dominate early succession in Antarctic seas, especially under the ice because of its nursery conditions. However, within a few weeks their dense populations may soak up much of the available silica and be largely replaced by another single-celled "plant" called a cryptophyte. Succession decrees that some forms of life are sitting on the bench waiting their turn when conditions change.

Another shared trait in both marine and terrestrial systems is ecological dominance – the ability of one species to muscle out other species and gain control over the rest of the community. When it does so, monopolizing the energy and nutrient pathways, its fate can influence the outcome of competitive struggles between whole ecosystems. The Antarctic Peninsula is one of the world's fastest-warming places.[30] Here, where melting ice is both increasing the temperature and decreasing the salinity of the sea water, diatoms have decreased significantly.[31] Cryptophytes are replacing them, even in early succession, in such prodigious numbers that they are capable of altering the reflectance of light.

The implications of this replacement are disturbing. Cryptophytes, being smaller than diatoms, are less optimum food for krill. As a consequence, krill have decreased by approximately 80 percent in some places.[32] Instead of diatoms feeding krill, which in turn feed most marine birds and mammals, cryptophytes feed barrel-shaped, free-floating blobs of gelatin called salps. While salps belong to the Phylum Chordata, as do we, salps put us in the shade reproductively. They reproduce rapidly by both sexual and asexual means and become hugely abundant very quickly,[33] to the point of sopping up all the daily primary production in some areas.[34] Most important, they are not a preferred food for organisms that occupy higher trophic levels, notably penguins, albatrosses, seals, and whales.[35] One of the first species to suffer appears to be Adélie penguins, which have declined significantly in most colonies along the Antarctic Peninsula. Adélies specialize heavily in krill.

———

Enthusiasm is a great way to build rapport in a confined place like a ship, and one of the tour leaders, Brian Keating, was a master of the art. Trained as a nature interpreter, he could make the most urbanely plumed tiny prion flying by the ship into the discovery of a lifetime. Even the uninterested who had never heard of a prion before wanted desperately to look through his binoculars and glimpse one as it hurtled by. At mealtimes, Brian would grandstand for the dining-room crowd, exclaiming over what we had seen since the last meal. On days when we had seen nothing but roiling waves, he would develop even the swells into objects of magnificence.

And they were. The sea-air interface is a gigantic membrane covering most of the surface of the Earth, its gas exchange key to survival of life. The ocean surface absorbs more carbon dioxide and releases more oxygen than the terrestrial world. The Southern Ocean is the only place on Earth without a land mass to impede a circling wind, and huge waves often pounded and crashed onto the *Sergey Vavilov*. Surface swells topped by frothy peaks broke over the bow with incredible force, one of the strongest on Earth. The restless blue-grey seascape spreading on all sides of the ship to the rolling horizon, the scudding clouds, the ever-changing play of light, all emphasized an otherworldliness.

Then, in sharp contrast were the perfectly calm, iceberg-strewn fjord waters around the Antarctic Peninsula itself. Part of the calmness was due to a studding of islands creating an inside passage along the west coast. But part was just our good luck, as the seas can be violent there, too. The iceberg show we experienced was a rich one; Antarctica's glaciers are calving more icebergs into the sea, now, due to climate warming. The bergs and surrounding waters were providing bed and breakfast for Adélie and chinstrap penguins, who entertained us passengers with their antics as the ship passed by.

Common, as well, to both terrestrial and marine ecosystems, happening everywhere on land and sea, is a plethora of opportunistic linkages among species. Natural selection makes it inevitable that species will find ways to benefit, often inadvertently, from the conditions that others set up. For example, many abundant species of marine phytoplankton produce a chemical, dimethyl sulphide, as they die, and they often die in large numbers, victims of altered sea conditions or viral attacks.[36] The chemical attracts krill that feed on the dying phytoplankton, but it also attracts krill predators, especially the abundant Wilson's storm petrel. The krill, being macerated by the feeding petrels, give off yet another odour that attracts even more petrels and other seabirds. Adding to the potpourri are larger seabirds, even albatrosses, lured by the presence of all the other birds. Soon the sea and air seethe with activity. Opportunism is a key to survival everywhere.

It is difficult, maybe impossible, to think of any ecosystem process that does not occur in both terrestrial and marine realms. The principles of populations are relevant, too. No species can escape them, especially the upper limit set by food. The diatom decline can be viewed as resulting from a lowering of environmental carrying capacity, here in a density-independent way (Chapter 8). The Adélie penguin decline was precipitated by a reduction of food. Similarly, numbers of Antarctic fur seals are depressed by lack of food in years of low krill production.[37]

The ways in which ecosystems may be either vulnerable or resilient are also similar. Ecosystems are structured the same way everywhere, with trophic levels, succession, predation, competition, population processes, and natural selection. For example, conferring resilience wherever found are alternate pathways for the flow of energy and nutrients made possible by species diversity. Conferring resilience, too, in changeable environments everywhere are generalist species.

Instead of describing the Southern Ocean, the ecological similarities of very different environments on Earth could be illustrated by looking at life around the hot, black, and anoxic deep-sea vents. There, species imported on the ocean currents have taken advantage of the resident chemosynthesizing bacteria and adapted, adjusted, and formed entire food webs of unique species: tubeworms up to two metres in length, limpets and mussels plastered to the submarine seamounts, giant clams, rat-sized amphipods, sea stars, and many others. Preying on these herbivores are omnivorous giant crabs, strangely shaped lobsters, predatory fish, octopi, snails. Then there are the pallbearers: scavenger shrimp, brittle stars, other fish. And finally the bacterial decomposers wait for the remaining crumbs.[38]

These similarities could be illustrated just as well by life in the physically demanding intertidal zone, where adaptations are required for alternating immersion in salt water and air, and for defence against predators coming from both worlds. Or, one could try to interpret the complex microbial interactions in open sea water. The life processes, the structures, and functions of all these ecosystems are the same.

Terrestrial and marine ecosystems are similar, too, in that both have fashioned species of great beauty – wood duck or wandering albatross, moose or minke whale – and constructed entire

ecosystems of great complexity – the southern Polar Front supporting everything from diatoms to penguins, or the shrub-steppe
semidesert extending from bunchgrasses to bighorns. All exist and
evolve under the same sun, with shared cycling nutrients, on the
same planet, spinning in the same solar system, in the same universe. All exhibit marked parallels in evolution.

One simple but profound conclusion flows from all these parallels, despite the extremely different physical conditions. It offers
a key ingredient in an evolutionary-ecological perspective. It provides a context for life. It is simply that the organization of life on
Earth is robust, that life support systems are based on inevitable
and enduring principles and processes, and that life, until our
planet passes out of the habitable zone, will persist. That is the
message of the twin and linked sciences of ecology and evolution
as understood today. It is a message of assurance.

But not necessarily a message of assurance for humans. It is the
life support processes that are inevitable and therefore enduring,
not their creations. The assembly line of life will keep turning out
new models as it has in the past – brontosaurs, diplodocus, giant
bison, humans – even though each model is destined for the scrap
heap. Extinction and replacement have always been, and will continue to be a vital part of life's adaptations to shifting conditions.
We humans are not immune, we who have already brought so
many near-calamities upon ourselves in our relatively short
residence.

Considering this robustness, what does the future hold? In a
worst-case scenario of widespread extinctions brought on by a
massive meteor strike over which we have no control, or even
ushered in by us in some nuclear disaster or climate-changing or
toxicant-releasing way, life could get pushed back to the forms that

persisted on Earth for 85 percent of its history – the unicellular. Extremophyles inform us that it would be difficult, even improbable, to push life back to complete extinction. Extremophyles would be the "comeback" organisms that natural selection would work on, once again. They have already jumped four of the six major hurdles that evolution has had to clear: replicating molecules, compartmentalized molecules (cells), chromosome structure, genetic code.[39] Making them was a huge accomplishment. We should not look down on these simple, single-celled beasts. Their only failings, from our lofty perspective, are the remaining two hurdles that they did not jump – multicellularity and language – but undoubtedly they perceive this issue differently.

Conceivably, and consolably, natural selection could make extremophyles adapt to conditions even more extreme than they have handled already, if necessary. The basic conditions they require seem reasonably secure: a liquid solvent in which chemical reactions can occur. Most biochemists agree that water is the only candidate because of its unique properties, such as the ability to participate in biochemical processes and, in its solid state, to float.[40] The various sources of energy for extremophyles are somewhat dependable, too: fermentation, chemosynthesis, proteorhodopsin, and even photosynthesis for cyanobacteria until we run out of carbon dioxide around 500,000-plus years from now. At that point, all terrestrial life will be extinct.

However, a disturbing aspect to life being pushed back that far is that evolution would have to jump, once again, one particularly high hurdle, that of creating multicellular life. We understand only poorly how that was accomplished some 700 million years ago, after almost 3 billion years of failure. It involved co-operation among some of those single cells floating in water that finally all pulled in the same direction to form nuclei, mitochondria, chloroplasts; it should not have been so difficult, nor have taken as long as it did. But the first two attempts formed lineages that died out; only the third attempt, about 100 million years later, took root.[41] Then it was as if a dam broke, and a flood of new species evolved to fill the Cambrian Seas. But because the step to multicellularism has proven so difficult on Earth, we look for signs of life on other planets with the expectation that the most we will find is unicellular life. Perhaps multicellularism anywhere involves chance at a level that is highly unlikely. After any mega-catastrophe on Earth, this step would be the one and only major obstacle to recovery to the full panoply of life that is the inevitable gift of natural selection.

Happily, this worst-case scenario of life reduced to single-celled organisms is the least probable scenario. Even after a worldwide mega-catastrophe, natural selection would likely have some surviving forms of multicellular life to work on. If so, then no major barrier would prevent the rich development of complex ecosystems once again.

And what are the possibilities of a percipient creature evolving again, a replacement for us? Could life be that resilient? Language is our great accomplishment. It fired our cultural evolution, and no species could fill our niche without it. Language – vocal communication with syntax and grammar – is unique to humans, far different from the simple signalling achieved by other species.[42] Language developed in small groups of hunter-gatherers about 300,000 to 200,000 years ago, driven by selection pressure for improved communication.[43] Its development was predicated by, or alternatively was possibly the result of, increased brain size, which occurred during the shift between genus *Australopithecus* and genus *Homo* by increased neotony (the prolonging of juvenile characteristics into later life).[44] Neotony extended gestation time, during which much brain development takes place in mammals, and then extended the period of parental dependence during which learning could occur.

Parallel to increased mental capacity of humans had to come anatomical changes in the location of the larynx lower in the throat, where air movement in the trachea would flow more forcefully through it. Genus *Homo* accomplished this adaptation, perhaps because of selection pressure on early inefficient sounds.[45]

Possibly a precursor to all this was the evolution of the hunter-gatherer lifestyle made possible by bipedalism. Walking upright allowed humans to use forelimbs in more creative ways than just locomotion.[46] There must have been much more to communicate once humans started making and using tools. That potential would have created selection pressure. And maybe a precursor to that was the adaptation of an opposable thumb, which is found in our earlier primate progenitors, and the development of manipulative lips and tongue. Thus, most likely, a chain of events, each link creating selection pressures for the next, had to occur to give us language and our resultant explosion of culture. Language, codified in writing, has allowed each generation to acquire the wisdom of the ages.

Why language has not happened in other lineages of social mammals – such as dogs, cats, mongooses, prairie dogs, ground squirrels – where one could envision similar adaptive advantages, is unclear. Perhaps it is the lack of chance mutations, or lack of initial evolutionary pressures for bipedalism to start things off. After all, what would prairie dogs talk about if they could talk? Already they can signal whether a predator is coming from the air or the ground, and signalling may be enough communication when all four feet must be used for locomotion. A lack of time for sophisticated communication to develop, however, is not the answer, as other mammalian lineages have been around just as long as primates.

This uncertainty leaves us unable to speculate whether we humans will ever have a speech-accomplished successor. If not, and if our own remarkable culture-driven adaptability fails to save us from catastrophe, then the last 5 billion years of the Earth's existence may be like the first 5 billion, with only the blip at the halfway point that is us. The beauty of a Neko Harbour, or a similar sculpture of nature, will still be there, even with nothing, once again, to appreciate it. Or does beauty not exist unless there is a beholder?

A happier prospect for life on Earth is for no catastrophe to occur, and for us and the species and ecosystems around us to go sailing on. Alas, the odds for this are poor. In only the first half of the planet's existence, it has experienced five major catastrophes, each causing a major extinction. And even now we are in the midst of perpetrating the sixth extinction.[47] We do not know how deeply we will cut life down. It could be deep, because we do not yet really take the damage we have inflicted seriously. We sign international "Biodiversity Accords" and ignore them; we publish a blue-ribbon "Millennium Ecosystem Assessment" that documents fully what needs to be done, and the world press gives it one day of coverage.[48] We . . .

But whether humans give the environment a licking or some errant meteor does, life is robust. Either way, it will rebound. After each of the five times it was knocked back so far, it has recovered to even greater diversity.[49] Recovery was inevitable because the construction crew was still around – the enduring processes of natural selection, order-for-free, energy capture, trophic structure, competition, predation, and a host of relationships within and between species. And, very importantly, multicellular species were still here with their genetic variability for natural selection to rework. Increasing adaptations to the environment were bound to reoccur.

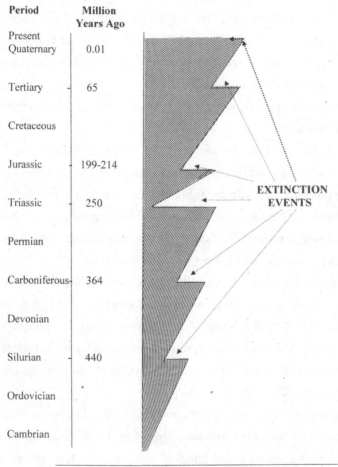

Period	Million Years Ago
Present Quaternary	0.01
Tertiary	65
Cretaceous	
Jurassic	199-214
Triassic	250
Permian	
Carboniferous	364
Devonian	
Silurian	440
Ordovician	
Cambrian	

EXTINCTION EVENTS

NUMBER OF LIVING GROUPS
(Modified from Primack, R.B., 1993. Conservation
Biology. Sinaur Associates, Sunderland, Massachusetts)

Even better, chances are good that the complement of future species that would do the repopulating would be remarkably similar to the species existing today. They have been similar in the past. Dinosaurs really are not so different from mammals. Just considering the forelimb, common between us are one bone in the upper arm, two bones in the forearm, a bunch of little bones in the wrist, long bones in the palm of the hand, smaller long bones down each finger. The rest of the skeletal plan is similarly equivalent. Also supporting the prediction of similarity in future species are the multitudinous examples of convergent evolution around us today, where very different classes, even phyla, of organisms, past and present, have evolved very similar structures, anatomies, and physiologies to deal with common problems. [50]

One response to life's capacity to endure is for us to just pillage the environment. We are! One-quarter of all the world's mammals are on the international red list of species threatened with extinction, along with 12 percent of all birds, 20 percent of reptiles, 32 percent of amphibians, and 34 percent of fish.[51] With every review, the percentages increase. If we doom these species, recovery, while inevitable, will take millions of years, as it has in the past. That is a long time to bestow an impoverished biological world on our possible descendants.

A better response to life's robustness is to demonstrate respect for life on Earth and its support systems, despite our macabre power over them.

It is a year later, now, and the Antarctic experience has mellowed in our minds. Dropped away has been anything trivial, like the feelings of seasickness during the below-deck lectures, or the lineups at the dining room when you couldn't stand up.

Like rendering a wood carving from a block of wood, those irrelevant chips have fallen away. What is left is the nub of the affair: the memories of the elegant flight of the albatross; the societies of penguins; the breaching whales and roiling seas.

Most salient is the fact that we have glimpsed first-hand, albeit artificially from a ship, both into the past and into the future. Here is where life began, and most probably where it will end.

But it won't end for a long while. And that is reason to celebrate.

CROUCHING ON THE HIGHEST STEP

FUELLING OUR PERSONAL, long-standing queries about how the natural world achieves self-order have been some dramatic experiences: back-to-back flamingos on Kenya's Lake Nakuru; masses of monarchs draping the oyamel firs in central Mexico; thundering clouds of magpie geese lifting off a northern Australian wetland. You cannot take spectacles like these for granted. They are triumphs of ecosystems and evolution.

Adding fuel, too, have been glimpses of species on the brink of extinction: a black stilt, the world's rarest shorebird, on a river bar in New Zealand, and half a dozen whooping cranes, the world's rarest waterbird, foraging in a Texas marsh. But in a less dramatic way, just as stimulating are our day-to-day encounters with the quiet processes of life support that pump along in the ponderosa pine – grassland ecosystem around our home, and the occasional glimpses of elk and moose, coyotes and cougars. What is the underlying commonality, the unifying concept?

The six years it has taken us to write this book have focused our efforts to address these questions. We have been able to juxtapose a rereading of our large collection of dog-eared field notebooks with an in-depth investigation of ecological literature, which has grown at an unprecedented rate. Gradually, a picture has taken shape, a synthesis of scattered facts about nature, coalescing and becoming more and more apparent, like film in an emulsion tray.

Life's self-organization happens through the actions of various novel processes that operate at each level of the biological hierarchy: species, populations, ecosystems. The CEO that keeps the entire enterprise on track is natural selection – the principal principle of life, that is, the primary source of order in living things. But natural selection does more than drive successful adaptations. It shapes the set of biotic processes that are its more ultimate and enduring achievement.

No particular alchemy is involved. Biotic processes are no more than the activities of living things, whether physiological or behavioural, and are just as much subject to selection as physical characteristics. Species that play out processes that work are favoured, and the processes with them; species that tinker with maladaptive processes, or don't adopt winning ones, fail, and the processes with them. How can it be otherwise? Living things poke and probe the possibilities like a twig in an eddy.

So, there *is* an overarching mechanism for the self-organization of life, and it is natural selection. But natural selection must have life to operate, so it needs, and indeed has, an accomplice – *ecosystems* –

the architects, builders, foundations, and castles of life. Forest or field, mountainside or marsh, prairie or pond, home range of mouse or moose, all are ecosystems. Boundaries are obscure, and scales differ, but all are self-ordered, real, living things.

Ecosystems are inevitable. They exhibit a type of adaptive immortality, that is, an ability to perpetuate themselves by changing with the times. But it is the *relationships and processes that are inevitable, not the immediate species.* Physical conditions change: continents drift; the planet wobbles; the sun intensifies; the atmosphere changes – and on a more immediate time scale, we humans mess things up. Yet faithfully, the same processes prevail, turning out new species. The reason is because the same solutions are repeatedly found by species living successively in the same functional niches and facing the same problems. Natural selection decrees this continuity of solutions.

No good analogy for an ecosystem exists on Earth, no manufacturing, robotic, or engineering equivalent. Its uniqueness lies in its winning trick for self-organization: *the products of ecosystems – species – go back and become part of the assembly line itself. In other words, ecosystems make their own components.* Like a child who finally masters bicycle riding, once an ecosystem gets going, it propels itself. Each component – insect, plant, wildlife – while adapting for its own well-being, is part of the selection pressure that moulds others and keeps them in place.

Thus, the self-order of ecosystems arises from a multitude of internal relationships. Some relationships are reciprocal: eat and be eaten; seek and provide habitat; use and provide energy and nutrients; photosynthesize and respire. Other relationships are structural, such as trophic levels and succession. Still others involve feedbacks, vital ones, like density dependence, that regulate populations. Yet others provide alternative pathways for energy and nutrients. Taken in aggregate, these relationships are a cat's cradle of cause and effect, the push-me-pull-you that endow ecosystems with the built-in redundancy and resilience that delivers their adaptive immortality.

The student of life still crouches on the highest step of the staircase of science, now a little higher because a few more steps have been hammered out. It is still impossible to see ahead, and there will always be room for speculation. But the student can look back down the staircase and see its outline more clearly, including the scientific support for the hypothesis that ecosystems are the very founts of life on Earth, that they are the architects, builders, foundations, and castles of life.

These thoughts are deeply satisfying; it is nice to know that life in some form will go on. The universe is an impersonal place, and while life here may not be unique given the multitude of other planets, it is the only inhabited place we will ever experience.

Moreover, nature appears more marvellous as a consequence of greater insights into its intricate mechanisms and support systems. An ecosystem-centred perspective on life also may make it easier for "a man's relationship to Nature to come very near to being personal," as Henry David Thoreau wrote. More marvellous, more personal, more beautiful, more worthy of respect.

But this all-encompassing view of ecosystems and life is not comforting in the short run. The biosphere *is* under assault. Climate change is happening faster than species can adapt. Ecosystems are under threat. Tropical forests are falling at the rate of a football field per minute. Amphibians worldwide are going extinct. Passerine birds are declining. The sixth great global extinction is well underway, with an alarming percentage of species slipping into the extinction vortex. Causing, driving, or exacerbating all these things is a current human population size and growth rate beyond any imaginable global, sustainable carrying capacity.

We are like prisoners lined up in front of a firing squad with many guns aimed at our heads. Just one environmental bullet can kill us. Alternative energy solutions, smokestack scrubbers, and other new technologies can perhaps deflect a few shots, but it is unlikely we can find, afford, and implement technical solutions to everything. And surprises keep cropping up. One hundred years ago, who would have even thought of acid rain, persistent toxic chemicals, antibiotic-resistant microbes, human-caused global climate change?

We need a way to call off the firing squad, a last-minute reprieve. Is there one? Could it spring from viewing ecosystems, indeed all nature, as more marvellous, more personal, more beautiful, more worthy of respect?

The answer could be yes if it triggered people to align themselves more closely with the enduring forces that underpin a sustainable Earth. The working end of such an alignment would be a rejection of the belief that growth is good, increased consumption is necessary, resources are unlimited, everything natural is stock-and-commodity just there for the taking, and human dominance over nature is a divine right. Replacing such a belief system would have to be a "deep ecology worldview" of just the opposite,[1] thereby tempering human activities in such ways as to maintain a healthy, species-rich biosphere – globally and everywhere locally.

Such a direction for human endeavours has been advocated for years. Thirty years ago, the World Conservation Strategy placed as its first objective, "to maintain essential ecological processes and life-support systems."[2] In 1982, the first principle of the United Nations' World Charter for Nature passed by the General Assembly stated that "Nature shall be respected and its essential processes shall not be impaired."[3] Similar international edicts have continued to appear every few years. Nonetheless, in 2005 the United Nations' Millennium Ecosystem Assessment stated that, "The challenge of reversing the degradation of ecosystems while meeting increasing demands for their services . . . involves significant changes in policies, institutions, and practices that *are not currently under way* [our emphasis]."[4]

In short, no shift in world-view has happened. The international edicts, stripped of any emotional connection or deeply felt respect for the intrinsic worth of nature, have simply failed to convince people or nations of the pressing need to tread more lightly on the Earth.

Perhaps that necessary emotional connection can be spawned *with a special conviction* if it is based on knowing as much as possible about its scientific underpinnings. Not just scientific facts, which are only the starting place, but science as a rational springboard for emotional connection – a passionate attachment.

Otherwise, we risk biological bankruptcy in the immediate future and foreclosure of the world as we know it.

Each spring, when the bitterroots are blooming on the mountainside around our Okanagan home, we take a week-long trip to the Klamath marshes on the California-Oregon border. This is a hotspot of biodiversity, a showcase of life. We go for the exhilaration of seeing, photographing, and recording the wealth of marsh birds in one of the outstanding wetlands of western North America.

We go, too, for the reassurance that life support processes are on duty, doing the best they can, despite humans. And they are on duty: energy flowing from cattails to hawks; nutrients moving from bottom sediments to eagles; competition functioning among territorial marsh wrens; predation happening when herons spear frogs; diseases like botulism and fowl pox lying in the mud ready to cull the weak; population processes measuring up and adjusting the densities of species; ecosystem linkages in operation. Orchestrating everything and assuring biological success for the best is the constant scrutiny of natural selection. The coordinated product? Beautiful creations of evolution – great egrets, black-crowned night herons, snow geese, cinnamon teal . . . and a functioning marsh machine.

At dusk we find some hollow where we can spend the night, because we are there, in part, to record the dawn chorus of birds. It starts before first light with a few restless coots, then a marsh wren or two. A heron wakes up and croaks before heading off for somewhere else. Some widgeon whistle. At daybreak a bittern begins his "pump-handle" call. Noisy flocks of ibises fly over, more marsh wrens rattle, and yellow-headed blackbirds punctuate the chorus. As the sun clears the cattails, birdsong ratchets up to a cacophony.

We are making the sound recordings for a practical reason, the same one that motivates us to record the dawn chorus at our cabin in Ontario, so that we can analyze it later for those frightening silent spaces, which, over the years, we have noted are increasing. But there is an even more immediate reason.

There is comfort in being immersed in biodiversity. Back in the recesses of our minds, if you take a long-term evolutionary perspective, is the consolation, even the peace, caused by certain knowledge of the inscrutability of life, the inevitability of ecosystem processes, "as incapable of being crushed or diverted from their ends as the rising tide."[5]

As the dawn touches the sky, we lie there and let the marsh sounds flood over us.

NOTES

PROLOGUE. A TWIG IN AN EDDY

1. Quammen, D. "Was Darwin wrong?" *National Geographic* November 2004.

CHAPTER 1: WHERE THE WOOD DUCK GOT ITS BEAUTY

1. A.M. Ross, *The birds of Canada: with descriptions of their plumage, habits, food, song, nests, eggs, times of arrival and departure* (Toronto: Rowsell and Hutchison, 1872).

2. E. Mayr, *What evolution is* (New York: Basic Books, 2001), 138.

3. D.W. Coltman, P. O'Donoghue, J.T. Jorgenson, J.T. Hogg, C. Strobeck, and M. Festa-Blanchet. "Undesirable evolutionary consequence of trophy hunting," *Nature* 426 (2003): 655–58.

4. Canadian Wildlife Service. "Wood duck," Ministry of the Environment, Ottawa, Ontario, 1991.

5. Mayr, 107–10.

6. Mayr, 280.

7. D.R. Wallace, *Beasts of Eden* (Berkeley: University of California Press, 2004), 7–17.

8. M. Balter. "Was Lamarck just a little bit right?" *Science* 8: 5463 (2000): 38.

9. Mayr, 204–6.

10. J.A. Coyne and H.A. Orr, *Speciation* (Sunderland, Massachusetts: Sinauer Associates, 2004), 420–21.

11. L.L. Shurtleff and C. Savage, *The wood duck and the mandarin* (Los Angeles: University of California Press, 1996).

12. C.S. Madsen, K.P. McHugh, and S.R. De Kloet. "A partial classification of waterfowl (Anatidae) based on single-copy DNA," *The Auk* 105 (1988): 452–59.

13. Various authors have noted difficulties in ageing using molecular clock data, and tendency towards underestimates, for example: J.C. Patton and J.C. Avise. "Evolutionary genetics of birds IV. Rates of protein divergence in waterfowl (*Anatidae*)," *Genetica* 698 (1986): 129–43. Also, M. Van Tuinen and S.B. Hedges. "The effects of external and internal fossil calibrations on the avian evolutionary timescale," *Journal of Palaeontology* 78 (2004): 45–50.

14. Many books explore this topic. An excellent overview is provided by: R. Lewin, *Complexity, life at the edge of chaos* (New York: Macmillan Publishing Company,1992). Also, S. Kauffman, *At home in the universe, the search for the laws of self-organization and complexity* (New York: Oxford University Press,1995).

15. G.G. Simpson, *The meaning of evolution* (New Haven: Yale University Press, 1949).

16. S. Camazine, J.L. Deneubourg, N.R. Franks, J. Sneyd, G. Theraulaz, and E. Bonabeau. *Self-organization in biological systems* (Princeton: Princeton University

Press, 2001), 12–13.

17. Lewin, 41.

18. D.C. Dennett, *Darwin's dangerous idea, evolution and the meanings of life* (New York: Simon and Schuster, 1995), 65.

19. D.T. Lindsay. "Simulating molluscan shell pigment lines and states: implications for pattern diversity," *Veliger* 24 (1977): 297–99.

20. Camazine, et al., 89.

21. Camazine, et al., 43.

CHAPTER 2: BIG ELK, LITTLE ELK

1. R.G. Thwaites, *Travels and explorations of the Jesuit Missionaries in New France 1610–1791*, vol. 72 (Cleveland: The Burrows Brothers Company, 1901), 216.

2. H.P. Biggar, *The works of Samuel De Champlain*, vol. 1 (Toronto: The Champlain Society. University of Toronto Press, 1922), 146.

3. Ernest Thompson Seton, *The lives of game animals*, vol. 3 (Garden City, New York: Doubleday, Doran and Company, 1909), 8.

4. Seton, 46.

5. J.W. Thomas and D.E. Toweill, *Elk of North America, ecology and management* (Harrisburg, Pennsylvania: Stackpole Books, 1982), 17.

6. V. Geist, *Deer of the world, their evolution, behavior, and ecology* (Mechanicsburg, Pennsylvania: Stackpole Books, 1998).

7. C.L. Williams, B. Lundrigan, and O.E. Rhodes, Jr. "Microsatellite DNA variation in Tule elk," *Journal of Wildlife Management* 68 (2004): 109–19. Also, E.P.M. Meredith, J.A.R. Odzen, J.D. Banks, R.S. Chaefer, H.B. Ernest, T.R.F. Amula, and B.P. May. "Microsatellite analysis of three subspecies of elk (*Cervus elaphus*) in California," *Journal of Mammalogy* 88 (2007): 801–8.

8. E. Jablonka and M.J. Lamb, *Evolution in four dimensions: genetic epigenetic, behavioral, and symbolic variation in the history of life* (Cambridge, Massachusetts: Massachusetts Institute of Technology, 2006).

9. E. Watters. "DNA is not destiny," *Discover Magazine* (November 2006): 33–37, 75.

10. N. McDevitt. "The nurture of things," *McGill Headway Magazine* vol. 2, no. 1., 2006 <http://www.mcgill.ca/headway>.

11. R.A. Waterland and R.L. Jirtle. "Transposable elements: targets for early nutritional effects on epigenetic gene regulation," *Molecular and Cellular Biology* 23 (2003): 5293–5300. This discovery followed from a trail of research over some years, such as H.D. Morgan, H.G.E. Sutherland, D.I.K. Martin, and E. Whitelaw. "Epigenetic inheritance at the agouti locus in the mouse," *Nature Genetics* 23 (1999): 314–18.

12. E. Pennisi. "Food, tobacco, and future generations," *Science* 310 (2005): 1760–61.

13. M.D. Anaway, A.S. Cupp, M. Uzumcu, M.K. Skinner. "Epigenetic transgenerational actions of endocrine disruptors and male fertility," Science 308 (June 3, 2005): 1466–69.

14. I.C.G. Weaver, N. Cervoni, F.A. Champagne, A.C.D. Alessio, S. Sharma, J.R. Seckl, S. Dymov, M. Szyf, and M. Meaney. "Epigenetic programming by maternal behaviour," *Nature Neuroscience* 7 (2004): 847–54.

15. D.K. Belyaev. "Destabilizing selection as a factor in domestication," *Journal of Heredity* 70 (1979): 301–8. (Cited by Jablonka and Lamb, 259–260.)

16. R. Lewin, *Complexity, life at the edge of chaos* (New York: Macmillan Publishing Company, 1992), 63–83.

17. Jablonka and Lamb, 273 and 358. This book provides a review of epigenetics and weaves it through other sources of inheritance from page 113 onward.

18. R.M. Nowak. "Another look at wolf taxonomy," in L.N. Carbyn, S.H. Fritts, and D.R. Seip, eds., *Ecology and conservation of wolves in a changing world*, occasional publication number 35 (Edmonton: Canadian Circumpolar Institute, 1995), 375–97.

19. R.K. Wayne and S.M. Jens. "Mitochondrial DNA analysis supports extensive hybridization on the endangered red wolf *(Canis rufus)*," *Nature* 351 (1991): 565–68.

20. P.J. Wilson, S. Grewal, I.D. Lawford, J.N.M. Heal, A.G. Granacki, D. Pennock, J.B. Theberge, M.T. Theberge, D.R. Voigt, W. Wadell, R.E. Chambers, P.C. Paquet, G. Goulet, D. Cluff, and B.N. White. "DNA profiles of the eastern Canadian wolf and the red wolf provide evidence for a common evolutionary history independent of the gray wolf," *Canadian Journal of Zoology* 78 (2000): 2156–66.

21. R.N. Nowak. "The original status of wolves in eastern North America," *Southeastern Naturalist* 1 (2002): 95–130.

22. Geist, 10.

23. C.D. Schlichting and M. Pigliucci. *Phenotypic evolution: a reaction norm perspective* (Sunderland, Massachusetts: Sinauer Associates), 68.

24. Jablonka and Lamb, 359.

25. J.B. Theberge and M.T. Theberge. *The wolves of Algonquin Park: A 12 year ecological study*, Publication number 56 (Waterloo, Ontario: Department of Geography, University of Waterloo, 2004).

26. Jablonka and Lamb, 1.

CHAPTER 3: DAWN CHORUS

1. R. Carson, *Silent spring* (New York: Houghton Mifflin, 1962).

2. J. Terborough, *Where have all the birds gone?* (Princeton: Princeton University Press, 1989).

3. B.J.M. Stutchbury, *Silence of the songbirds* (Toronto: HarperCollins Canada, 2007).

4. J. Klicka and R.M. Zink. "The importance of recent ice ages in speciation: a failed paradigm," *Science* (1997) 277: 1666–69.

5. D.A. Sibley, *The Sibley guide to birds* (New York: Alfred A. Knopf, 2000), 349.

6. D. Stap, *Bird song* (New York: Scriber, 2005), 84. Also, D. Kroodsma, *The singing life of birds* (Boston: Houghton Mifflin Company, 2005), 89.

7. Kroodsma, 44–55.

8. K.C.R. Kerr, M.Y. Stoeckle, C.J. Dove, L.A. Weigt, C.M. Francis, and P.N. Hebert. "Comprehensive DNA barcode coverage of North American birds," *Molecular Ecology Notes* 7 (2007): 535–43.

9. P. Marler and P. Tamura. "Song dialects in three populations of white-crowned sparrows," *Condor* (1962) 64: 368–77.

10. Reviewed by L.F. Baptista, "Nature and nurturing in avian vocal development," in D.E. Kroodsma and E.H. Miller, *Ecology and evolution of acoustic communication in birds* (Ithaca, New York: Cornell University Press, 1996), 39–59.

11. D.E. Kroodsma. "Song learning, dialects, and dispersal in the Bewick's wren," *Zeitschrift fur Tierpsychologie* 35 (1974): 352–80.

12. H.E. Howard, *Territory in bird life* (London: John Murray, 1920).

13. J.B. Falls. "Properties of bird song eliciting responses from territorial males," in C.G. Sibley, ed., *Proceedings of the XIIIth International Ornithological Congress* (Lawrence, Kansas: Allen Press, 1963), 259–71.

14. S. Emlen. "An experimental analysis of the parameters of bird song eliciting species recognition," *Behaviour* 41 (1972): 130–71.

15. S.F. Lovell and M.R. Lein. "Neighbor-stranger discrimination by song in a suboscine bird, the alder flycatcher, *Empidonax alnorum*," *Behavioral Ecology* 15 (2004): 799–804.

16. Reviewed by P.K. Stoddard. "Vocal recognition of neighbors by territorial passerines," in Kroodsma and Miller, *Ecology and evolution of acoustic communication in birds*, 339–74.

17. Reviewed by J.B. Falls. "Individual recognition by sounds in birds," in D.E. Kroodsma and E.H. Miller, eds., *Acoustic communication in birds*, vol. 2 (New York: Academic Press, 1982), 237–78.

18. D.E. Kroodsma. "The ecology of avian vocal learning," *Bioscience* 33 (1983): 165–71.

19. E.S. Morton. "Ecological sources of selection on avian sounds," *The American Naturalist* 109 (1975): 17–34.

20. M.J.M. Martens. "Foliage as a low-pass filter: experiments with model forests in an anechoic chamber," *Journal of the Acoustic Society of America* 67 (1980): 66–72.

21. D.G. Richards and R.H. Wiley. "Reverberations and amplitude fluctuations in the propagation of sound in a forest: implications for animal communications," *The American Naturalist* 115 (1980): 381–99.

22. D. Stap, 88.

23. D. Kroodsma, *The singing life of birds*, 79.

24. C.H. Greenewalt, "How birds sing," in B.W. Wilson, *Birds: readings from Scientific American* (San Francisco: W.H. Freeman and Company, 1980), 228–39.

25. D. Kroodsma, *The singing life of birds*, 355–56.

26. This subject is reviewed thoroughly in R.G. Foster and L. Kreitzman. *Rhythms of life, the biological clocks that control daily lives of every living thing* (New Haven: Yale University Press, 2004), 66–152. Also, E.A. Brenowitz and D.E. Kroodsma, "The

neuroethology of birdsong," in Kroodsma and Miller, *Ecology and evolution of acoustic communication in birds*, 285–304.

27. Foster and Kreitzman, 152.

28. E. Jablonka and M.J. Lamb, *Evolution in four dimensions; genetic, epigenetic, behavioral, and symbolic variation in the history of life* (Cambridge, Massachusetts: Massachusetts Institute of Technology, 2006), 161.

29. P. Marler and S. Peters. "Developmental overproduction and selective attrition: new processes in the epigenesist of birdsong," *Developmental Psychobiology* 15 (1982): 369–78.

30. Jablonka and Lamb, 290.

31. Jablonka and Lamb, 360.

32. C.S. Robbins, J.R. Sauer, R.S. Greenberg, and S. Droege. "Population declines in North American birds that migrate to the neotropics," *Proceedings of the National Academy of Sciences* 86 (1989): 7658–62. Also, J. Tuxill. "Losing strands in the web of life: vertebrate declines and the conservation of biological diversity," *Worldwatch Paper* no. 141 (Washington, D.C.: Worldwatch Institute, 1998). Also, H. Youth. "Winged messengers, the decline of birds," *Worldwatch Paper* no. 165 (Washington, D.C.: Worldwatch Institute, 2003).

33. S. Weidensaul, *Living on the wind, across the hemisphere with migratory birds* (New York: North Point Press, 1999), 369–70.

34. J. Muir, in Linnie Marsh Wolfe, ed., *John of the mountains: the unpublished journals of John Muir* (Madison: University of Wisconsin Press, 1938).

CHAPTER 4: THE PERFECT MOOSE

1. P.S. Martin and H.E. Wright, Jr. "Pleistocene extinctions," *Proceedings of the VIIth Congress of the International Association for Quaternary Research*, vol. 6 (New Haven: Yale University Press, 1967).

2. Martin and Wright, Jr., 95.

3. R.M. Nowak, *North American Quaternary Canis*. Museum of Natural History, monograph no. 6 (Lawrence, Kansas: University of Kansas Press, 1979), 110.

4. E. Mayr. "Speciational evolution or punctuated equilibria," in A. Somit and S. Peterson, *The dynamics of evolution* (New York: Cornell University Press, 1992), 21–48.

5. F. Darwin, in John Murray, publisher, *The life of Charles Darwin*, reprint of 1902 publication (London: Senate Press, 1995). Quote: "Believing as I do that man in the distant future will be a more perfect creature than he is now." R.C. Berwick, review of *Climbing Mount Improbable* by R. Dawkins: <http://bostonreview.net>.

6. National Book Company, *The world's greatest court trial* (Cincinnati, Ohio, 1925).

7. S. Wright, "The roles of mutation, inbreeding, crossbreeding, and selection in evolution," *Proceedings of the Sixth International Congress on Genetics*, 1 (1932): 356–66.

8. L. Van Valen. "A new evolutionary law," *Evolutionary Theory* 1 (1973): 1–30. Also, S.A. Kauffman. "Escaping the red queen effect," *The McKinsey Quarterly* 1 (1995): 118–29.

9. T. Haines, *Walking with dinosaurs, a natural history* (New York: Dorling Kindersley Publishing, 2000), 261.

10. A.L. Wallace, *The geographical distribution of mammals*, vol. 1 (London: Macmillan, 1876).

11. T. Haines, *Walking with prehistoric beasts, a prehistoric safari.* (New York: Dorling Kindersley Publishing, 2001), 37.

12. E. Mayr, *What makes biology unique?* (New York: Cambridge University Press, 2004), 121.

13. N. Eldredge, *The patterns of evolution* (New York: W.H. Freeman and Company, 2000), 63.

14. V. Geist, *Deer of the world, their evolution, behaviour, and ecology* (Mechanicsburg, Pennsylvania: Stackpole Books, 1998), 235. Further references on body and antler size in moose are also from Geist, 7–17, 228–32, 350.

15. R.L. Peterson, *North American moose* (Toronto: University of Toronto Press, 1978).

16. C. Darwin, *The descent of man, and sexual selection in relation to sex*, vol. 2 (Princeton: Princeton University Press, 1871).

17. Kauffman.

18. E. Mayr, *What evolution is* (New York: Basic Books, 2001), 143.

19. T. Hobbes, *Leviathan, or the matter, forme, and power of a commonwealth, ecclesiasticall and civill*, 1651.

20. J.B. Theberge and M.T. Theberge. *Wolf country, eleven years tracking Algonquin wolves* (Toronto: McClelland & Stewart, 1998), 238.

21. L.A. Gavrilov and N.S. Gavrilova. "Evolutionary theories of aging and longevity," *The Scientific World Journal* 2 (2002): 339–56.

22. Mayr, *What evolution is*, 142.

23. Gavrilov and Gavrilova.

24. Theberge and Theberge, 104–5.

25. R.C. Anderson. "The ecological relationships of meningeal worm and native cervids in North America," *Journal of Wildlife Diseases* 8 (1972): 304–10.

26. Mayr, *What evolution is*, 140–43.

27. Mayr, *What makes biology unique?*, 112, 155.

28. S.J. Gould, 1981. "Evolution as fact and theory," in *Hen's teeth and horse's toes* (New York: W.W. Norton and Company, 1981), 253–62.

CHAPTER 5: GIVE ME LAND, LOTS OF LAND

1. J.C. Theberge. "Grizzly bear resource selection and scale: scale-dependent selection of resource characteristics and landscape pattern by female grizzly bears in the eastern slopes of the Canadian Rocky Mountains," Ph.D. thesis, 2002, University of Calgary, Calgary, Alberta.

2. J.C. Theberge. "Resource selection by female grizzly bears," in *Final report of the Eastern Slopes Grizzly Bear project* (Calgary: University of Calgary Press, 2005), 173.

3. S. Stevens and M. Gibeau. "Home range analysis," in *Final report of the Eastern Slopes Grizzly Bear project*, 144.

4. B.L. Horejsi and B.K. Gilbert. "Conservation of grizzly bear populations and habitat in the northern Great Bear Rainforest," *Biodiversity* 7 (2006): 3–10.

5. R. M. Nowak, *Mammals of the world*, vol. 2 (Baltimore: John Hopkins University Press, 1991), 1090.

6. R.E. Grumbine, *Ghost bears – exploring the biodiversity crisis* (Washington D.C.: Island Press, 1992), 66.

7. E.O. Wilson, *Sociobiology, the new synthesis* (Cambridge, Massachusetts: Belknap Press of Harvard University Press, 1975), 268–69.

8. B.J.M.Stutchbury and E.S. Morton. *Behavioral ecology of tropical birds* (San Diego: Academic Press, 2001), 62.

9. R.L. Smith, *Ecology and field biology*, 5th ed. (New York: Harper Collins, 1996), 418.

10. Wilson, 271–72.

11. P. Lurtz, P.J. Garson, and L.A. Wauters. "Effects of temporal and spatial variations in food supply on the space and habitat use of red squirrels," *Journal of the Zoological Society of London* 251 (2000): 167–78.

12. M.A. Mares, T.E. Lacher, M.R. Willig, N.A. Bitar, R. Adams, A. Klinger,and D. Tazik. "An experimental analysis of social spacing in *Tamias striatus*," *Ecology* 63 (1982): 267–73.

13. T.M. Smith, and H.H. Shugart. "Territory size variation in the ovenbird," *Ecology* 68 (1987): 695–704.

14. F.L. Carpenter. "Food abundance and territoriality: to defend or not to defend?" *American Zoologist* 27 (1987): 387–99.

15. D. Kroodsma, *The singing life of birds, the art and science of listening to birdsong* (Boston: Houghton Mifflin, 2005), 74.

16. Kroodsma, 352.

17. J.W. Terborough, 1980. "The conservation status of neotropical migrants: present and future," in A. Keast and E.S. Morton, eds., *Migrant birds in the Neotropics: ecology, behavior, distribution and conservation* (Washington, D.C.: Smithsonian Institute, 1980).

18. J.C. Kricher, *A neotropical companion: an introduction to the animals, plants, and ecosystems of the new world tropics*, 1st ed. (Princeton: Princeton University Press, 1989), 130–31.

19. S. Hilty, *Birds of tropical America* (Shelburne, Vermont: Chapters Publishers, 1994), 281.

20. Hilty, 36.

21. Stutchbury and Morton, 64–65.

22. Hilty, 278–81.

23. Stutchbury and Morton, 81.

24. J.H. Rappole and D.W. Warner, "Ecological aspects of migrant bird behavior in Vera

Cruz, Mexico," in A. Keast and E.S. Morton, eds., *Migrant birds in the Neotropics,* 353–93.

25. P. Schwartz. "Some considerations on migratory birds," in Keast and Morton, eds., *Migrant birds in the Neotropics,* 31–34.

26. R.E. Ricklefs. "Birds of two worlds: temperate-tropical migration systems," *Trends in Ecology and Evolution* 17 (2002): 302–3. Also Schwartz, 31–34.

27. P.P. Marra. "The role of behavioral dominance in structuring patterns of habitat occupancy in a migrant bird during the non-breeding season," *Behavioral Ecology* 11 (2000): 299–308.

28. J.M. Bates. "Winter territory behavior of gray vireos," *Wilson Bulletin* 104 (1992): 425–33. Also Rappole and Warner.

29. E.S. Morton and B.J.M. Stutchbury. "The significance of mating system and non-breeding behavior to population and forest patch use by migrant birds," *United States Department of Agriculture, Forest Service. General Technical Report PSW-GTR-191,* 2005.

30. E.O. Willis. "Competitive exclusion and birds at fruiting trees in western Columbia," *The Auk* 83 (1966): 479–80.

31. Hilty, 106.

32. R.E. Kenward. "Hawks and doves: attack success in goshawk flights at wood-pigeons," *Journal of Animal Ecology* 47 (1978): 449–60.

33. P. Colinvaux, *Ecology* (New York: John Wiley and Sons, 1986), 293.

34. Colinvaux, 282–84.

35. A.T. Bergerud and M.W. Gratson. *Adaptive strategies in population ecology of northern grouse,* vol. II (Minneapolis: University of Minnesota Press, 1988), 465–72.

36. J.C. Kricher, "Neotropical birds," Chapter 12 in *A Neotropical companion: an introduction to the animals, plants, and ecosystems of the new world tropics,* 2nd ed. (Princeton: Princeton University Press, 1997), 279.

CHAPTER 6: HIGH-STAKES LIVING

1. The primary sources for the attraction of the historic Platte River for cranes are: P. A. Johnsburg, *Crane music, a natural history of American cranes* (Lincoln: Bison Books, University of Nebraska Press, 1998), 41–56, 93. Also, P. Matthiesen, *Birds of heaven* (New York: Farrar Straus Giroux, 2001), 260–65. Also, R.B. Primack, *Essentials of conservation biology* (Sunderland, Massachusetts: Sinauer Associates, Inc., 1993), 312.

2. J.C. Finlay, *A bird finding guide to Canada* (Toronto: McClelland & Stewart, 1984), 26.

3. H. P. Biggar, *The voyages of Jacques Cartier,* no. 11 (Ottawa: Publications of the Public Archives of Canada, 1924), 198.

4. The birds of North America Online, Cornell Lab of Ornithology. < http://www.bna.birds.Cornell.edu/bna>. Also <http://www. pacificflyway.gov>. Also, S. Weidensaul, *Living on the wind* (New York: North Point Press, 1999), 225.

5. <http://www. pacificflyway.gov>.

6. The birds of North America Online.

7. *Discover* magazine, vol. 27 (2006), no. 1: 43.

8. R.G. Foster, L. Kreittzman. *Rhythms of life* (New Haven: Yale University Press, 2004), 135. Also, R.D. Estes, *The safari companion* (White River Junction, Vermont: Chelsea Green Publishing, 1993), 121.

9. Estes, 120.

10. M. E. Soule, *Conservation biology* (Sunderland, Massachusetts: Sinauer Associates, Inc., 1986), 363.

CHAPTER 7: TEN THOUSAND REINDEER

1. A. Leopold, *Game management* (New York: Charles Scribner's Sons, 1933), 50.

2. J.G. Jorgensen, *Oil age Eskimos* (Berkeley: University of California Press, 1990).

3. E. Gruening, *The State of Alaska* (New York: Random House, 1954), 96.

4. S. Jackson, *Introduction of domestic reindeer into Alaska.* 10th annual report (Washington, D.C.: Government Printing Office, 1900), 17.

5. S. Jackson, *Introduction of domestic reindeer into Alaska.* 13th annual report (Washington, D.C.: Government Printing Office, 1903), 13.

6. S. Jackson, *Introduction of domestic reindeer into Alaska.* 11th annual report (Washington, D.C.: Government Printing Office, 1901), 15.

7. S. Jackson, *Introduction of domestic reindeer into Alaska.* 14th annual report (Washington, D.C.: Government Printing Office, 1904), 21.

8. S. Jackson, *Introduction of domestic reindeer into Alaska.* 13th annual report, 79.

9. S. Jackson, *Introduction of domestic reindeer into Alaska.* 15th annual report (Washington, D.C.: Government Printing Office, 1905), 65.

10. Anonymous. "Appendix D: community profiles" in *Bering Straits coastal resource service area, coastal management plan* (Anchorage: Department of Natural Resources, State of Alaska, 2006).
    <http://www.dnr.alaska.gov/coastal/. . . /FinalPlans/BeringStraits>.

11. E.J. Kormondy, *Concepts of ecology*, third ed. (Englewood Cliffs, New Jersey: Prentice-Hall, 1984), 93.

12. R.A. Rausch. "On the land mammals of St. Lawrence Island, Alaska," *The Murrelet* 34 (1953): 18–26.

13. D.R. Klein. "The introduction, increase, and crash of reindeer on St. Matthew Island," *Journal of Wildlife Management* 32 (1968): 350–67.

14. V.B. Scheffer. "The rise and fall of a reindeer herd," *Scientific Monthly* 73 (1951): 356–62.

15. Jorgensen.

16. J. Parkes, K. Tustin, and L. Stanley. "The history and control of red deer in the Takahe area, Murchison Mountains Fiordland National Park," *New Zealand Journal of Ecology* 1 (1978): 145–52.

17. P. Mahoney, *History of animal pest control, historic huts identification study.* New

Zealand Department of Conservation, 2000.
<http://www.doc.govt.nz/upload/Conservation/History-Animal-Pest-Control.pdf>

18. J. Diamond, *Collapse, how societies choose to fail or succeed* (New York: Penguin Books, 2005), 91.

19. Diamond, 121–31.

20. L.B. Keith, *Wildlife's ten-year cycle* (Madison: The University of Wisconsin Press, 1963).

21. Keith. Also, H. Viljugrein, O .Lingjaerde, N. Stenseth, and M.S. Boyce. "Spacio-temporal patterns of mink and muskrat in Canada during a quarter century," *Journal of Animal Ecology* 70 (2001): 671–82.

22. C.S. Elton, *Animal ecology* (London: Sidgwick and Jackson Limited, 1927). Also, C.S. Elton. "Periodic fluctuations in the number of animals: their causes and effects," *British Journal of Experimental Biology* 2 (1924): 119–63.

23. A.R.E. Sinclair, J.M. Gosline, G. Holdsworth, C.J. Krebs, S. Boutin, J.N.M. Smith, R. Boonstra, and M. Dale. "Can the solar cycle and climate synchronize the snow-shoe hare cycle in Canada? Evidence from tree rings and ice cores," *The American Naturalist* 141 (1993): 173–98.

24. E. Ranta, , J. Lindstrom, V. Kaitala, H. Kokko, H. Linden, and E. Helle. "Solar activity and hare dynamics: a cross-continental comparison," *The American Naturalist* 149 (1997): 765–75.

25. J.B. Theberge. "Evaluation of the winter range of white-tailed deer in Point Pelee National Park," *The Canadian Field-Naturalist* 92 (1978): 19–23.

26. J.B. Theberge and D.A. Gauthier. "Factors influencing densities of territorial male ruffed grouse, Algonquin Park, Ontario," *Journal of Wildlife Management* 46 (1982): 263–68.

27. J.B. Theberge and M.T. Theberge, *The wolves of Algonquin Park, a 12 year ecological study*. Publication series no. 56 (Waterloo: Department of Geography, University of Waterloo, 2004).

28. A.J. Nicholson. "The balance of animal populations," *Journal of Animal Ecology* 2 (1933): 132–78.

29. H.G. Andrewartha and L.C. Birch. *The distribution and abundance of animals* (Chicago: University of Chicago Press, 1954).

30. Leopold, 49.

CHAPTER 8: LABRADOR WILD

1. I. Newton. "The role of food in limiting bird numbers," *Ardea* 68 (1980): 11–30.

2. J.B. Theberge and D.A. Gauthier. "Models of wolf-ungulate relationships: when is wolf control justified?" *Wildlife Society Bulletin* 13 (1985): 449–58.

3. D.R. Klein. "The introduction, increase, and crash of reindeer on St. Matthew Island," *Journal of Wildlife Management* 32 (1968): 350–67.

4. Cited by D.L. Allen, *Our wildlife legacy* (New York: Funk and Wagnall Company,

1954), 235.

5. L.L. Snyder, *Arctic birds of Canada* (Toronto: University of Toronto Press, 1957), 97, 240.

6. L.B. Keith. "Role of food in hare population cycles," *Oikos* 40 (1983): 385–95. Also, L.B. Keith. "Dynamics of snowshoe hare populations," in H.H. Genowats, ed., *Current mammalogy* (New York: Plenum Press, 1990), 119–95.

7. C. Rohner, F.I. Doyle, and J.N. Smith, "Great horned owls," in C.J. Krebs, S. Boutin, R. Boonstra, eds., *Ecosystem dynamics of the boreal forest, the Kluane project* (New York: Oxford University Press, 2001), 339–76.

8. R.O. Skoog. "Ecology of the caribou (*Rangifer tarandus granti*) in Alaska," Ph.D. thesis, University of California, Berkeley, California, 1968. Also, F.L. Miller, R.H. Russell, and A. Gunn, *Peary caribou and muskoxen on western Queen Elizabeth Islands, N.W.T. 1972–1974.* Canadian Wildlife Service, Report series no. 40, 1977.

9. C.S. Houston. "Spread and disappearance of the greater prairie chicken (*Tympanuchus cupido*) on the Canadian prairies and adjacent areas," *The Canadian Field-Naturalist* 116 (2002): 1–21.

10. J.N.M. Smith and N.F.G. Folkard. "Other herbivores and small predators," in Krebs, et al., eds., *Ecosystem dynamics of the boreal forest* (2001): 261–72.

11. K.E. Hodges, C.J. Krebs, D.S. Hik, C.I. Stefan, E.A. Gillis and C.E. Doyle, "Snowshoe hare demography," in Krebs, et al., eds., *Ecosystem dynamics of the boreal forest*, 141–78.

12. D. Wallace, *The lure of the Labrador wild* (New York: Fleming H. Revell Company, 1905).

13. D.R. Seip. "Factors limiting woodland caribou populations and their interrelationships with wolves and moose in southeastern British Columbia," *Canadian Journal of Zoology* 70 (1992): 1494–1503.

14. A.W.F. Banfield, *The mammals of Canada* (Toronto: University of Toronto Press, 1974), 397.

15. A. Fyvie, W.G. Ross, and N.A. Labzoffsky. "Tularemia among beaver and muskrat in Ontario," *Canadian Journal of Medical Science* 30 (1952): 250–55.

16. J.N. Pauli, S.W. Buskirk, E.S. Williams, and W.H. Edwards. "A plague epizootic in the black-tailed prairie dog (*Cynomys ludovicianus*)," *Journal of Wildlife Diseases* 42 (2006): 74–80.

17. E.G. Bolen and W.L. Robinson. *Wildlife ecology and management*, 3rd ed. (Englewood Cliffs, New Jersey: Prentice Hall, 1995), 131.

18. W. Wishart. "Bighorn sheep," in J.L. Schmidt and D.L. Gilbert, eds., *Big game of North America, ecology and management* (Harrisburg, Pennsylvania: Stackpole Books, 1978), 161–71.

19. M.D. Samuel. "Wetlands, waterfowl, and avian cholera outbreaks," United States Geological Survey, National Wildlife Health Centre, Information sheet, 2002.

20. Bolen and Robinson, 136.

21. J.S. Sachs. "Super bugged," *Discover Magazine* (March 2008): 58–62.

22. V. Geist. "Moose," in *Deer of the world, their evolution, behaviour, and ecology* (Harrisburg, Pennsylvania: Stackpole Books, 1998), 223–54.

23. Bolen and Robinson, 126.

24. J.B. Theberge and M.T. Theberge, *Wolf country, eleven years tracking Algonquin wolves* (Toronto: McClelland & Stewart, 1998).

25. A.T. Bergerud. "Caribou," in Schmidt and Gilbert, eds., *Big game of North America*, 83–102.

26. J.P. Kelsall, *The migratory barren-ground caribou of Canada* (Ottawa: Queen's Printer, 1968).

27. G.M. Happ, H.J. Hudson, B. Kimberlee, and L.J. Kennedy. "Prion protein genes in caribou from Alaska," *Journal of Wildlife Diseases* 43 (2007): 224–28.

28. Results of this research are found in the following documents: W.K. Brown. "The ecology of a woodland caribou herd in central Labrador," M.Sc. thesis, Department of Biology, University of Waterloo, Waterloo, Ontario, 1986. Also, W.K. Brown and J.B. Theberge. "The effects of extreme snowcover on feeding-site selection by woodland caribou," *Journal of Wildlife Management* 54 (1990): 161–68; W.K. Brown, and J.B. Theberge. "The calving distribution and calving-area fidelity of a woodland caribou herd in central Labrador," 2nd North American caribou workshop (October 1984), McGill University, McGill Subarctic Research Paper series no. 40 (1985): 57–67.

29. C.J. Krebs, S. Boutin, R. Boonstra, A.R.E. Sinclair, J.N.M. Smith, M.R.T. Dale, and K.M.R. Turkington. "Impact of food and predation on the snowshoe hare cycle," *Science* 269 (1995): 1112–15. Also, A.R.E. Sinclair, J.M. Fryxell, and G. Caughley. "Chapter 12: Consumer-Resource dynamics," in *Wildlife ecology, conservation, and management* (Oxford: Blackwell, 2006), 214–15.

30. J.A. Schaefer, A.M. Veitch, F.H. Harrington, W.K. Brown, J.B. Theberge, and S.N. Luttich. "Demography of decline of the Red Wine Mountains caribou herd," *Journal of Wildlife Management* 63 (1999): 580–87.

31. <http://www.airtraining.forces.gc.ca/training/fmt/goosebay>.

CHAPTER 9: ALL INDIVIDUALS ARE NOT CREATED EQUAL

1. P.R. Ehrlich, *The population bomb* (New York: Ballantine Books, 1968).

2. D.H. Meadows, D.L. Meadows, J. Randers, and W.W. Behrens III, *The limits to growth, a report for the Club of Rome's project on the predicament of mankind* (New York: Signet, Mentor, and Plume Books, 1972).

3. D. Chitty. "The natural selection of self-regulatory behaviour in animal populations," *Proceedings of the Ecological Society of Australia* 2 (1967): 51–78.

4. D. Chitty, *Do lemmings commit suicide? Beautiful hypothesis and ugly facts* (New York: Oxford University Press, 1996).

5. A. Watson and R. Moss. 1970. "Dominance, spacing behaviour and aggression in relation to population limitation in vertebrates," in A. Watson, ed., *Animal populations in*

relation to their food resources (Oxford: Blackwell, 1970), 167–218.

6. R.L. Smith and T.M. Smith. *Elements of ecology* 4th ed. (San Francisco: Benjamin/Cummings Science Publishing, 1998), 173.

7. J. J. Christian. "The adrenal-pituitary system and population cycles in mammals," *Journal of Mammalogy* 31 (1950): 274–359. Also, J. J. Christian and D.E. Davis. "Adrenal glands in female voles (*Microtus pennsylvanicus*) as related to reproduction and population size," *Journal of Mammalogy* 47 (1966): 1–18.

8. V.C. Wynne-Edwards, *Animal dispersion in relation to social behaviour* (Edinburgh: Oliver and Boyd, 1962).

9. W.D. Hamilton. "The genetical theory of social behaviour," *Journal of Theoretical Biology* 7 (1964): 1–52. Topic reviewed by E.O. Wilson, *Sociobiology, the new synthesis* (Cambridge, Massachusetts: The Belnap Press of Harvard University Press, 1975), 117–20.

10. J.B. Theberge and J.F. Bendell. "Differences in survival and behaviour of rock ptarmigan (*Lagopus mutus*) chicks among years in Alaska," *Canadian Journal of Zoology* 58 (1980): 1638–42.

11. R.B. Weeden and J.B. Theberge. "The dynamics of a fluctuating population of rock ptarmigan in Alaska," *Proceedings of the XVth International Ornithological Congress,* The Hague, Holland (1972), 90–106.

12. M.R. Werbach. "Nutritional influences on aggressive behaviour," *Journal of Orthomolecular Medicine* 7 (1995): 45–51.

13. A. Watson and R. Moss. "A current model of population dynamics in red grouse," *Proceedings of the XVth International Ornithological Congress,* The Hague, Holland (1972), 134–49.

14. J.D. Schaechter and R.J. Wurtman. "Serotinin release varies with brain tryptophan levels," *Brain Research* 532 (1990): 293–310. Also, M.F. Bear, B.W. Connors, and M.A. Pardiso. *Neuroscience: exploring the brain* (Philadelphia: Lippincott Williams and Wilkins, 2006), 581.

15. E.R. Radwanski and R.L. Last. "Tryptophan biosynthesis and metabolism: biochemical and molecular genetics," *Plant Cell* 7 (1995): 921–34.

16. V. Selas, O. Hogstad, S. Kobro, and T. Rafoss. "Can sunspot activity and ultraviolet-B radiation explain cyclic outbreaks of forest moth pest species?" *Proceedings of the Royal Society, Biological Sciences* 271 (2004): 1897–1901.

17. Werbach.

18. Jablonka and Lamb. *Evolution in four dimensions: genetic, epigenetic, behavioural, and symbolic variation in the history of life* (Cambridge, Massachusetts: Massachusetts Institute of Technology, 2006), 144.

19. C.H.D. Clarke. "Fluctuations in the numbers of ruffed grouse, *Bonasa umbellus* (Linne), with special reference to Ontario," *University of Toronto Studies,* Biological Series no. 41, 1936.

20. J.B. Theberge and D.A. Gauthier. "Factors influencing densities of territorial male

ruffed grouse, Algonquin Park, Ontario," *Journal of Wildlife Management* 46 (1982): 263–68.

21. Jablonka and Lamb, 285–93.

22. J.F. Bendell and K.J. Szuba. "Balance and levels in populations of spruce grouse *Dendragopus canadensis*," *Proceedings of the International Union of Game Biologists XXIst Congress*, 2 (1993): 114–19.

23. R.L. Smith, *Ecology and field biology*, 5th ed. (New York: HarperCollins College Publishers, 1996), 393.

24. A.J. Nicholson. "The balance of animal populations," *Journal of Animal Ecology* 2 (1933): 132–78.

25. P.L. Errington, *Of predation and life* (Ames: Iowa State University Press, 1967).

26. C.J. Krebs, *Ecology, the experimental analysis of distribution and abundance* (New York: Harper and Row, 1972), 340.

27. A variety of web sources were used, including: United Nations Population Information Network <www.un.org/popin>; World Bank Report *Beyond Economic Growth* www.worldbank.org; International Data Base of the United States Census Bureau <www.census.gov/opc/www/idb/worldpopinfo.html>; Population Reference Bureau <www.prb.org>; International Institute for Applied Systems Analysis <www.iiasa.ac.at>.

28. E.O. Wilson, *The future of life* (New York: Alfred A. Knopf, 2002), 33.

29. United Nations *Millennium Ecosystem Assessment* (United Nations: 2005).

CHAPTER 10: WAR ON THE SHRUB-STEPPE

1. D. Nierenberg. "Happier meals, rethinking the global meat industry," *Worldwatch Paper* 171 (Washington, D.C.: Worldwatch Institute, 2005), 8.

2. A.H. Benton and W.E. Werner, Jr., *Field biology and ecology* (New York: McGraw-Hill Publishers, 1966), 103.

3. R.L. Smith, *Ecology and field biology*, 5th ed. (New York: HarperCollins Publishers, 1996), 227.

4. D. Alt, *Glacial Lake Missoula and its humongous floods* (Missoula, Montana: Mountain Press, 2001).

5. S. Trimble, *The sagebrush ocean* (Reno: University of Nevada Press, 1999).

6. *Washington Post*, July 9, 2004, from webpage: <http://www.washingtonpost.com>.

7. <http://usgovinfo.about.com/library/weekly/aa102497.htm>.

8. <http://www.pnl.gov/ecology/shrub/ALE/ALE>.

9. C. Darwin, *The origin of species* (Edison, New Jersey: Castle Books, 2004 edition), 85.

10. R. Law and A.R. Watkinson. "Competition," in J.M. Cherrett, ed., *Ecological concepts: the contribution of ecology to an understanding of the natural world; 29th symposium of the British Ecological Society* (Oxford: Blackwell Scientific Publications, 1989), 272.

11. C.R. Townsend, M. Begon, and J.L. Harper, *Essentials of ecology* (Malden, Massachusetts: Blackwell Publishing, 2002), 204.

12. The birds of North America Online, Cornell Lab of Ornithology and American Ornithologists' Union <http://www.bna.birds.Cornell.edu/bna>.

13. Townsend, Begon, and Harper, 204.

14. H.N. Mozingo, *Shrubs of the Great Basin, a natural history* (Reno: University of Nevada Press, 1987), 276.

15. P.R. Cutright, *Lewis and Clarke: pioneering naturalists* (Lincoln: University of Nebraska Press, 1969), cited in A.S. Laliberte and W.J. Ripple. "Wildlife encounters by Lewis and Clark: a spatial analysis of interactions between native Americans and wildlife," *Bioscience* 53 (2003): 994–1003.

16. E.E. Rich, ed., *The Hudson's Bay Company 1670–1870*, vol. II 1763–1870 (London: The Hudson's Bay Record Society, vol. 22, 1959), 591.

17. E.E. Rich, ed., *Peter Skeene Ogden's Snake Country journals, 1824–25 and 1825–26* (London: The Hudson's Bay Record Society, vol. 13, 1950), 259.

18. F. Egan, *Fremont, explorer for a restless nation* (Reno: University of Nevada Press, 1985), 253.

19. Laliberte and Ripple. Also, D. Osborne. "Archaeological occurrences of pronghorn antelope, bison, and horse in the Columbia Plateau," *Scientific Monthly* (November 1953), 260–69.

20. A.L. Haines, ed., *Journal of a trapper: Russell Osborne* (Lincoln: University of Nebraska Press, 1965), cited in T.M. Bonnicksen, *America's ancient forests* (New York: John Wiley and Sons, 2000), 85.

21. E.T. Seton, *Lives of game animals*, vol. III, 2nd edition (New York: Doubleday, Doran and Company, 1927).

22. C.W. Johnson. "Protein as a factor in the distribution of the American bison," *Geographical Review* 41 (1951): 330–31.

23. C.L. Wilson and W.E. Loomis, *Botany* (New York: The Dryden Press, 1958), 81.

24. R. Daubenmire. "The western limits of the range of the American bison," *Ecology* 66 (1985): 622–24.

25. P.S. Martin and C. Szuter. "War zones and game sinks in Lewis and Clark's west," *Conservation Biology* 13 (1997): 36–45.

26. E.T. Seton, *Lives of game animals*, vol. IV, 2nd edition (New York: Doubleday, Doran and Company, 1927), 280.

27. R.N. Mack and J.N. Thompson. "Evolution in steppe with few large hoofed mammals," *The American Naturalist* 119 (1982): 757–73.

28. There are many references to this topic. Two are cited, the first relative to southern British Columbia, the second an authoritative overview: E.W. Tisdale. "The grasslands of the southern interior of British Columbia," *Ecology* 28 (1947): 346–82; V.E. Shelford, *The ecology of North America* (Urbana: University of Illinois Press, 1963), 350–51.

29. Mozingo, 280.

30. C.J. Krebs, *The experimental analysis of distribution and abundance* (New York: Harper

and Row Publishers, 1972), 312.

31. G.W. Cox, *Alien species and evolution* (Washington, D.C.: Island Press, 2004), 55.

32. L. Brown, *Grasslands. The Audubon Society Nature Guides* (New York: Alfred A. Knopf, 1985), 79–80.

33. Krebs, 313.

34. Mack and Thompson.

35. Smith, 488. Also, Law and Watkinson, 264.

36. R. Manning, *Grassland, the history, biology, politics, and promise of the American prairie* (New York: Penguin Books, 1995), 180.

37. J.O. Klemmedson and J.G. Smith. "Cheatgrass (*Bromus tectorum* L.)," *Botanical Review* 30 (1964): 226–62.

38. Manning, 178.

39. Krebs, 313.

40. L. Leopold, ed., *Round River, from the journals of Aldo Leopold* (Oxford, New York: Oxford University Press, 1953), 165.

41. United Nations list of protected areas. Web address:
    <http://www.iucn.org/wpc2003/english/outputs.un.html>

CHAPTER 11: BEHIND THE SCENES

1. S.R. Severs and J.B. Theberge. *Oiseau Bay, Pukaskwa National Park: An assessment of potential environmental alterations* (Cornwall, Ontario: Parks Canada, 1975).

2. S.T.A. Pickett and P.S. White. *The ecology of natural disturbance and patch dynamics* (San Diego: Academic Press, 1985), 18.

3. Yukon environmentally significant area series, 9 volumes with various authors: J.B. Theberge, J.G. Nelson, M. Fitzsimmons, M. Stabb, D. Sauchyn, J.D. Bastedo, B. Hans Bastedo, M.A. Sauchyn, K. O'Reilly (Waterloo, Ontario: President's Committee on Northern Studies, University of Waterloo, 1986 and 1987).

4. L. Berger, R. Speare, P. Daszak, D.E. Green, A.A. Cunningham, C.L. Goggin, R. Slocombe, M.A. Ragan, A.D. Hyatt, K.R. McDonald, H.B. Hines, K.R. Lips, G. Marantelli, and H. Parkes. "Chytridiomycosis causes amphibian mortality associated with population declines in the rainforests of Australia and Central America," *Proceedings of the National Academy of Sciences* 95 (1998): 9031–36.

CHAPTER 12: TEAM PLAY

1. J.M. Cherrett, "Key concepts: the results of a survey of our members' opinions," in J.M. Charrett, ed., *Ecological concepts, the contribution of ecology to an understanding of the natural world* (London: Blackwell Scientific Publications, 1989), 1–16.

2. Discussed in the introduction of many ecology texts, such as, R.L. Smith, *Ecology and field biology* (New York: HarperCollins, 1996).

3. D.F. Tombach. "Dispersal of whitebark pine seeds by Clarke's nutcracker: a mutualistic hypothesis," *Journal of Animal Ecology* 51 (1982): 451–67.

4. Smith, 580–94. All ecology books have chapters describing mutualistic and symbiotic relationships.

5. .C. Luvall and H.R. Holbo. "Measurements of short term thermal responses of coniferous forest canopies using thermal scanner data," *Remote Sensing the Environment* 27 (1989): 1–10.

6. G. Hardin, *Filters against folly* (New York: Penguin Books, 1985), 55–59.

7. A detailed description of early events for life is provided by A.H. Knoll, *Life on a young planet, the first three billion years of evolution on Earth* (Princeton: Princeton University Press, 2003), 81–110.

8. J.E. Lovelock, *Gaia, a new look at life on Earth* (New York: Oxford University Press, 1987), 31.

9. J. Cowper Powys, *The art of happiness* (New York: Simon and Shuster, 1935), 33–55.

10. Classic papers are: E.P. Odum. "Trends expected in stressed ecosystems," *Bioscience* 35 (1985), 419–422; D.J. Rapport. "What constitutes ecosystem health?" *Perspective in biology and medicine* 33 (1989): 120–133; D.J. Schaeffer, E.E. Herricks, and H.W. Kerster. "Ecosystem health: I. Measuring ecosystem health," *Environmental Management* 12 (1988): 445–55.

11. H. Seyle, *The story of the adaptation syndrome* (Montreal: Acta Incorporated, 1952).

12. R.M. Nowak, *Walker's mammals of the world,* fifth edition (Baltimore: The John Hopkins University Press, 1991), 1428–29.

13. A.J. Whitten, S.J. Damanik, J. Anwar, and N. Hisyam. *The ecology of Sumatra* (Yogyakarta, Indonesia: Gadjah Mada University Press, 1984), 386.

14. Nowak, 200-4.

CHAPTER 13: OUT OF THE SLAG HEAP OF NATURE

1. T. Flannery, *The eternal frontier, an ecological history of North America and its peoples* (New York: Grove Press, 2001), 29.

2. The sources for the prehistoric species mentioned here are: K. Blount and M. Crowley, *Dinosaur encyclopaedia, from dinosaurs to the dawn of man.* American Museum of Natural History (New York: Dorling-Kindersley Publishing, 2001). Also, T. Haines, *Walking with dinosaurs, a natural history* (New York: Dorling-Kindersley Publishing, 2000); Also, T. Haines, *Walking with prehistoric beasts, a prehistoric safari* (New York: Dorling-Kindersley Publishing, 2001), 73.

3. R.J. Davidson. "Foundation of the land, bedrock geology," in J.B. Theberge, ed., *Legacy, the natural history of Ontario* (Toronto: McClelland & Stewart, 1989), 36–48.

4. A.H. Richardson, *A report on the Ganaraska watershed* (Toronto: Dominion and Ontario Governments, 1944). Quotes in this paragraph are from pages 91, 90, and 94 respectively.

5. N. Mika and H. Mika, *Encyclopedia of Ontario, vol. II, Place names in Ontario.* (Belleville, Ontario: Mika Publishing Company, 1977), 655.

6. Richardson, frontispiece.

7. J.M. Cherrett, *Ecological concepts* (London: Blackwell Scientific Publications, 1989).

8. P. Berton, *Klondike* (Toronto: McClelland & Stewart, 1958), 293.

9. C.E. Richmond, D.L. Breitburg, and K.A. Rose. "The role of environmental generalist species in ecosystem functioning," *Ecological Modelling* 188 (2005): 279–95. Also, E.P. Odum, *Basic ecology* (New York: Saunders College Publishing, 1983), 362.

10. R. Dawkins, *The ancestor's tale* (London: Orion Books Limited, 2004), 232–34.

11. S. Markey. "Alien possums gobbling New Zealand forests," *National Geographic News* (April 25, 2006) Also, D.A. Wardle, G.M. Barker, G.W. Yeates, K.I. Bonner, and A. Ghani. "Introduced browsing mammals in New Zealand natural forests: aboveground and belowground consequences," *Ecological Monographs* 71 (2001): 587–614.

12. T.T. Veblen and G.H. Stewart. "The effects of introduced wild animals on New Zealand forests," *Annals of the Association of American Geographers* 72 (1982): 372.

13. G. Hardin, 1985. *Filters against folly* (New York: Penguin Books, 1985), 55–59.

14. Odum, 358. Also, R.L. Smith, *Ecology and field biology*, 5th ed. (New York: HarperCollins Publishers, 1996), 448.

15. R. Dawkins, *The selfish gene* (Oxford: Oxford University Press, 1976).

16. R.A. Miller, *Genes and the agents of life* (Cambridge, England: Cambridge University Press, 2005), 188.

17. D. Sloan Wilson. "Introduction: multilevel selection theory comes of age," *The American Naturalist* 150 (1997): S1-S4.

18. D. Sloan Wilson. "Biological communities as functionally organized units," *Ecology* 78 (1997): 2018–24.

19. D. Sloan Wilson and W. Swenson. "Community genetics and community selection," *Ecology* 84 (2003): 586–588.

20. W. Swenson, D. Sloan Wilson, and R. Elias. "Artificial ecosystem selection," *Proceedings of the National Academy of Sciences of the United States of America* 97 (2000): 9110–14.

21. D. Sloan Wilson, "Introduction: multilevel selection theory comes of age."

22. D. Sloan Wilson. "Altruism and organism: disentangling the themes of multilevel selection theory," *The American Naturalist* 150 (1997): S122-S134.

23. R. Cropp and A. Gabric. "Ecosystem adaptation: do ecosystems maximize resilience?" *Ecology* 83 (2002): 2019–26.

24. S.T.A. Pickett and P.S. White, *The ecology of natural disturbance and patch dynamics* (San Diego: Academic Press, 1985), 18.

25. T.M. Bonnicksen, *America's ancient forest, from ice age to discovery* (New York: John Wiley and Sons, 2000), 223.

26. C.F. Steiner, Z.T. Long, J.A. Krumins, and P.J. Morin. "Population and community resilience in multitrophic communities," *Ecology* 87 (2006): 996–1007.

CHAPTER 14: THE PTARMIGAN'S DILEMMA

1. J.B. Theberge and G.C. West. "Significance of brooding to the energy demands of Alaskan rock ptarmigan chicks," *Arctic* 26 (1973): 138–48.

2. N. Lane, *Oxygen, the molecule that made the world* 1st paperback edition (Oxford: Oxford University Press, 2003), 19.

3. A.H. Knoll, *Life on a young planet, the first three billion years of evolution on Earth* (Princeton: Princeton University Press, 2003), 21.

4. The chemistry of chlorophyll is exceedingly complex. Summaries about it and its role in photosynthesis are provided in most ecology texts, such as R.L. Smith, *Ecology and field biology*, 5th ed. (New York: HarperCollins, 1996), 154–59. More depth is found in A. Tresbst and M. Avron, eds., *Photosynthesis* (New York: I. Springer-Verlag, 1977). Much ongoing research is published in the scientific *Journal of Photosynthesis Research*.

5. G. Wald. "Fitness in the universe: choices and necessities," *Origins of Life* 5 (1974): 7–27. Also, S.C. Morris, *Life's solutions, inevitable humans in a lonely universe* (Cambridge, England: Cambridge University Press, 2003), 106–11.

6. Morris, 108.

7. S. Franck, W. von Bloh, and C. Bounama. "Determination of habitable zones in extra-solar planetary systems: where are Gaia's sisters?" *Journal of Geophysical Research* 105 (2000), no. E1: 1651–58. Also, S. Franck, A. Block, W. von Bloh, C. Bournama, H.-J. Schellnhuber, and Y. Svirezhev. "Habitable zone for Earth-like planets in the solar system," *Planetary and Space Science* 48 (2000): 1099–1105; S. Franck, A. Block, W. von Bloh, C. Bounama, I. Garrido, and H.-J. Schellnhuber. "Planetary habitability: is Earth commonplace in the Milky Way?" *Naturwissenschaften* 88 (2001): 416–26.

8. P. Ball, *Life's matrix: a biography of water* (New York: Farrar, Straus and Giroux, 1999).

9. Lane, 48.

10. P.F. Hoffman and D.P. Schrag. "Snowball Earth," *Scientific American* 282 (2000): 68–75.

11. Knoll, 208–15.

12. T. Flannery, *The eternal frontier: an ecological history of North America and its peoples* (London: William Heinemann, 2001), 148.

13. G.M. Hewitt. "Some genetic consequences of ice ages, and their role in divergence and speciation," *Biological Journal of the Linnean Society* 58 (1996): 247–76. Also E.C. Pilelou, *After the ice ages* (Chicago: The University of Chicago Press, 1991).

14. R.S. Bradley, *Quaternary Paleoclimatology: methods of paleoclimate reconstruction* (Boston: Allen and Unwin, 1985).

15. J.E. Francis. "A 50-million-year-old fossil forest from Strathcona Fiord, Ellesmere Island, Arctic Canada: evidence for a warm polar climate," *Arctic*, 41 (1988): 314–18.

16. Morris, 109.

17. J.B. Graham, R. Dudley, N.M. Aguilar, and C. Gans. "Implications of the late Paleozoic oxygen pulse for physiology and evolution," *Nature* 375 (1995): 117–20. Cited in T.R.E. Southwood, *The story of life* (Oxford: Oxford University Press, 2003), 90.

18. J.R. Petit, J. Jouzel, D. Raynaud, N.I. Barkov, J.-M. Barnola, I. Basile, M. Benders, J. Chappellaz, M. Davis, G. Delaygue, M. Delmotte, V.M. Kotlyakov, M. Legrand, V.Y. Lipenkov, C. Lorius, L. Pepin, C. Ritz, E. Saltzman, and M. Stievenard. "Climate and atmospheric history of the past 420,000 years from the Vostok ice core, Antarctica," *Nature* 399 (1999): 429–36. Data graphed by United Nations Environmental Program / GRID-Arendal, Vital Climate Graphics.

19. A.J. Gaston, D.K. Cairns, R.D. Elliot, and D.G. Noble. "A natural history of Digges Sound," *Canadian Wildlife Service Report* series no. 46 (Ottawa, Ontario, 1981).

20. .B. Theberge and C.H.R. Wedeles. "Prey selection and habitat partitioning in sympatric coyote and red fox populations, southwest Yukon," *Canadian Journal of Zoology* 67 (1989): 1285–90.

21. J.B. Theberge and T.J. Cottrell. "Food habits of wolves in Kluane National Park," *Arctic* 30 (1979): 189–91. Also, D.A. Gauthier and J.B. Theberge. "Wolf predation in the Burwash caribou herd, southwest Yukon," *Rangifer* 1 (1986): 137–44.

22. J.H. Brown, *Macroecology* (Chicago: The University of Chicago Press, 1995), 124.

23. Smith, 369.

24. D.L. Murray and S. Larivie. "The relationship between foot size of wild canids and regional snow conditions: evidence for selection against a high footload?" *Journal of the Zoological Society of London* 256 (2002): 289–99.

25. R.F. Johnston and R.K. Selander. "House sparrows: rapid evolution of races in North America," *Science* 144 (1964): 548–50.

26. Initially, various authors proposed that the Pleistocene glaciations played major roles in avian speciation: J.C. Avise, D. Walker, and G.C. Johns. "Speciation durations and Pleistocene effects on vertebrate phylogeography," *Proceedings of the Royal Society of London* 265 (1998): 1707–12. This conclusion appeared inconsistent with molecular clock data by J. Klicka and R.M. Zink. "The importance of recent ice ages in speciation: a failed paradigm," *Science* 277(1997): 1666–69. Debate that ensued includes: J. Klicka and R.M. Zink. "Pleistocene effects on North American songbird evolution." *Proceedings of the Royal Society of London* 266 (1999): 695–700. Also, B.S. Aborgast and J.B. Slowinski. "Pleistocene speciation and the mitochondrial DNA clock, Technical comment." *Science* 282 (1998).

27. S. Franck, A. Block, W. von Bloh, C. Bounama, I. Garrido, and H.-J. Schellnhuber. "Planetary habitability: is Earth commonplace in the Milky Way?" *Naturwissenschaften* 88 (2001): 416–26.

28. J. Lovelock, *The revenge of Gaia* (London: Penguin Books, 2006), 64.

29. T.M. Lenton and W. von Bloh. "Biotic feedback extends the life span of the biosphere," *Geophysical Research Letters* 28 (2001): 1715–18.

30. Lovelock, 45–46.

31. Theberge and West.

CHAPTER 15: LET THERE BE LIFE

1. R. Carson, *The edge of the sea*. First Mariner Book Edition (New York: Houghton Mifflin Company, 1998), 250.

2. V. Smetacek and S. Nicol. "Polar ocean ecosystems in a changing world," *Nature* 437 (2005): 362–68.

3. National Marine Fisheries Service, Office of Protected Resources, United States Department of Commerce: <http://www.nmfs.noaa.gov/pr/species/mammals /cetaceans/spermwhale.htm>.

4. K.E. Slack, C.M. Jones, T. Ando, G.L. Harrison, R.E. Fordyce, U. Arnason, and D. Penny. "Early penguin fossils, plus mitochondrial genomes, calibrate avian evolution," *Molecular Biology and Evolution* 23 (2006): 1144–55.

5. P.G. Davis. "The oldest record of the genus Diomedea, *Diomedea tanakai* sp. nov. (Procellariiformes: Diomedeidae): an albatross from the Miocene of Japan," *Bulletin of the Natural History Museum of London*, Series C. 29 (2003): 39–48.

6. G.A. Nevitt. "Olfactory foraging by Antarctic Procellariiform seabirds: life at high Reynolds numbers," *Biological Bulletin* 198 (2000): 245–53. Also, G.A. Nevitt and F. Bonadonna. "Seeing the world through the nose of a bird: new developments in the sensory ecology of procellariiformes seabirds," *Marine Ecology Progress Series* 287 (2005): 292–95.

7. R.W. Van Buskirk and G.A. Nevitt. "Evolutionary arguments for olfactory behaviour in modern birds," *Australian Association for Chemosensory Science* 10 (2007): 1–6.

8. Van Buskirk and Nevitt.

9. R.C. Murphy, *Logbook for Grace, whaling brig Daisy* 1912–1913, reprinted 1965 (Chicago: Time Life Books), 148.

10. Species profile and threats database, *Diomedea exulans – Wandering Albatross*. Department of Environment, Water, Heritage and the Arts, Australia: <http://www.environment.gov.au>.

11. P. C. Withers. "Aerodynamics and hydrodynamics of the 'hovering' flight of Wilson's storm petrel," *Journal of Experimental Biology* 80 (1979): 83–91.

12. J. Sarmiento, N. Gruber, M.A. Brzezinski, and J.P. Dunne. "High-latitude controls of the thermocline nutrients and low latitude biological productivity," *Nature* 427 (2003): 56–60.

13. J.C. Venter, K. Remington, J.F. Heidelberg, A.L. Halpern, D. Rusch, J.A. Eisen, D.Y. Wu, I. Paulsen, K.E. Nelson, W. Nelson, D.E. Fouts, S. Levy, A.H. Knap, M.W. Lomas, K. Nealson, O. White, J. Peterson, J. Hoffman, R. Parsons, H. Baden-Tillson, C. Pfannkoch, Y. Rogers, H.O. Smith. "Environmental genome shotgun sequencing of the Sargasso Sea," *Science* 304 (2004): 66–74. Also, J.C. Venter, *A life decoded, my genome, my life* (New York: Penguin Group, 2007), 344.

14. J. Copley. "All at sea," *Nature* 415 (2002): 572–74.

15. J. Montoya, C.M. Holl, J.P. Zehr, A. Hansen, T.A. Villareal, and D.G. Capone. "High rates of N_2-fixation by unicellular diazotrophs in the oligotrophic Pacific," *Nature*

430 (2004): 1027–32.

16. D. Karl, and M. Karner. "Archaeal dominance in the mesopelagic zone of the Pacific Ocean," *Nature* 409 (2001): 507–10. Also, Copley.

17. Karl and Karner. Also, M.L. Coleman, M.B. Sullivan, A.C. Martiny, C. Steglich, K. Barry, E.F. DeLong, and S.W. Chisholm. "Genomic islands and the ecology and evolution of Prochlorococcus," *Science* (2006): 1768–70.

18. D. Werner. "Productivity studies on diatom cultures," *Helgoland Marine Research* 20 (1970): 97–103.

19. C.L. Van Dover, *Deep-ocean journeys, discovering new life at the bottom of the sea* (Reading, Massachusetts: Addison-Wesley, 1996), 56.

20. R.B. Hoover, E.V. Pikuta, A.K. Bej, D. Marsic, W.B. Whitman, J. Tang, and P. Krader. "*Spirochaeta americana* sp. nov., a new haloalkaliphilic, obligately anaerobic spirochete isolated from soda Mono Lake in California," *International Journal of Systematics and Evolutionary Microbiology* 53 (2003): 815–21.

21. B.R. Kelemen, M. Du, and R.B. Jensen, "Proteorhodopsin in living color: diversity of spectral properties within living bacterial cells," *Biochimica et Biophysica Acta* 1618 (2003): 25–32.

22. J. Nianzhi, F. Fuying, and W. Bo. "Proteorhodopsin – a new path for biological utilization of light energy in the sea," *Chinese Science Bulletin* 51 (2006): 889–96. Also, L. Gomez-Consarnau, J.M. Gonzalez, M. Coll-Llado, P. Gourdon, T. Pascher, R. Neutze, C. Pedros-Alio, J. Pinhassi. "Light stimulates growth of proteorhodopsin-containing marine flavobacteria," *Nature* 445 (2007): 210–13; Also, Venter, 346.

23. O. Beja, A.L. Koonin, E.V. Suzuki, M.T. Hadd, A. Nguyen, L.P. Jovanovich, S.B. Gates, C.M. Feldman, R.A. Spudich, J.L. Spudich, and E.F. DeLong. "Bacterial rhodopsin: evidence for a new type of phototrophy in the sea," *Science* 289 (2000): 1902–4.

24. O. Beja. "Light-sensing protein illuminates sun-loving ocean bacteria," *Public Library of Science Biology* 3 (2005): 287.

25. A.H. Knoll, *Life on a young planet, the first three billion years of evolution on Earth* (Princeton: Princeton University Press, 2003), 88.

26. Knoll, 103.

27. N. Lane, *Oxygen, the molecule that made the world*, paperback edition (Oxford: Oxford University Press, 2003), 24.

28. D.N. Thomas, *Frozen oceans, the frozen world of pack ice* (Buffalo: Firefly Books, 2004), 111.

29. Thomas, 118.

30. A. Atkinson, V. Siegel, E. Pakhomov, and P. Rothery. "Long-term decline in krill stock and increase in salps within the Southern Ocean," *Nature* 432 (2004): 100–103.

31. M.A. Moline and B.B. Prezelin. "Long-term monitoring and analyses of physical factors regulating variability in coastal Antarctic phytoplankton biomass, in situ

productivity and taxonomic composition over subseasonal, seasonal and interannual time scales," *Marine Ecology Progress Series* 145 (1996): 143–160.

32. M. Allsopp, R. Page, P. Johnston, and D. Santillo, *Oceans in peril: protecting marine biodiversity* (Washington, D.C.: Worldwatch Institute, 2007), 23.

33. V. Smetacek and S. Nicol. "Polar ocean ecosystems in a changing world," *Nature* 437 (2005): 362–68.

34. R. Perissinotto and E.V. Pakhomov. "The trophic role of the tunicate *Salpa thompsoni* in the Antarctic marine ecosystem," *Journal of Marine Science* 17 (1998): 361–74. Also, C.D. Dubischar, and U.V. Bathmann. "Grazing impacts of copepods and salps on phytoplankton in the Atlantic sector of the southern ocean," *Deep Sea Research Part II: Topical Studies in Oceanography* 44 (1997): 415–433.

35. M.A. Moline, H. Claustre, T.K. Frazer, O. Schofields, and M. Vernet. "Alteration of the food web along the Antarctic Peninsula in response to a regional warming trend," *Global Change Biology* 10 (2004): 1973–80. Also A. Atkinson, V. Siegel, E. Pakhomov, and P. Rothery. "Long-term decline in krill stock and increase in salps within the southern ocean," *Nature* 432: 100–103.

36. Copley.

37. Proposal to de-list Antarctic fur seals as specially protected species. Antarctic Treaty Consultative Meeting, 2006: <http://www.atcm2006.gov.uk>.

38. Van Dover.

39. E. Szathmary and J. Maynard Smith. "Major evolutionary transitions," *Nature* 374 (1995): 227–32.

40. P. Ball, *Life's matrix: a biography of water* (New York: Farrar, Straus and Giroux, 1999). Also, G. Wald. "Fitness in the universe: choices and necessities," *Origins of Life* 5 (1974): 7–27.

41. S. J. Gould, *Wonderful life, the Burgess Shale and the nature of history* (New York: W.W. Norton and Company, 1989), 58–61.

42. E. Mayr, *What evolution is* (New York: Basic Books, 2001), 253.

43. E. Mayr, *This is biology, the science of the living world* (Cambridge, Massachusetts: Harvard University Press, 1997), 240.

44. S.J. Gould, *The Panada's thumb, more reflections in natural history* (New York: W.W. Norton and Company, 1980), 106–7.

45. R. Dawkins, *Unweaving the rainbow, science, delusion and the appetite for wonder* (New York: Houghton Mifflin Company, 1998), 294–97.

46. R. Leakey and R. Lewin, *Origins reconsidered, in search of what makes us human* (London: Little, Brown and Company, 1992), 92.

47. R. Leaky and R. Lewin, *The sixth extinction, patterns of life and the future of humankind* (New York: Doubleday, 1995).

48. R. Hassen, R. Scholes, and N. Ash, eds., *Ecosystems and human well-being, current state and trends.* Volume 1. *Millennium Ecosystem Assessment* (Washington, D.C.: Island Press, 2005).

49. R.B. Primack, *Essentials of conservation biology* (Sunderland, Massachusetts: Sinauer Associates Incorporated, 1993), 78.

50. S.C. Morris, *Life's solution, inevitable humans in a lonely universe* (New York: Cambridge University Press, 2003).

51. J. Baillie, L.A. Bennun, C. Hilton-Taylor, T.M. Brooks, and S.N. Stuart, *IUCN red list of threatened species: a global species assessment* (Gland, Switzerland: World Conservation Union, Species Survival Commission, 2004).

EPILOGUE: CROUCHING ON THE HIGHEST STEP ·

1. B. Devall and G. Sessions, *Deep ecology, living as if nature mattered* (Salt Lake City: Peregrine Smith Books, 1985).

2. *World Conservation Strategy.* (Gland, Switzerland: International Union for the Conservation of Nature, United Nations Environment Program, World Wildlife Fund, 1980).

3. *World Charter of Nature.* United Nations General Assembly, 37th Session, Resolution 37/7, 1982.

4. R. Hassen, R. Scholes, and N. Ash, eds., *Ecosystems and human well-being, current state and trends. Volume 1. Millennium Ecosystem Assessment* (Washington, D.C.: Island Press, 2005).

5. R. Carson, *The edge of the sea*, First Mariner Book Edition (New York: Houghton Mifflin Company, 1998), 250.

ACKNOWLEDGEMENTS

TO THE EXTENT THAT THIS BOOK represents a reflection across more than thirty years of field research, conservation activities, and teaching, it is impossible to acknowledge all those who contributed. Underpinning much of our work was the University of Waterloo in Waterloo, Ontario – its administrators, colleagues, and students – who made the academic environment what it should be: a place of thought-and-idea entrepreneurship and learning. We thank all those people, including those in agencies who financially supported our endeavours.

We also thank our daughters and their spouses for critical review of the manuscript, as well as our publishers, McClelland & Stewart, especially Dinah Forbes, whose structural and detailed advice enabled us to better thread our story.

Paul Eby

John and Mary Theberge have spent more than thirty years conducting field research in the USA, Canada, and overseas. They have collaborated on numerous scientific and popular articles and in 1994 were jointly awarded the Equinox Citation for Environmental Achievement. John Theberge was, until his recent retirement, a professor of ecology and conservation biology in the faculty of environmental studies at the University of Waterloo, where he taught since 1970. Mary Theberge is an educator and wildlife researcher and has presented many popular programs about their discoveries. They are the authors of several previous books, including *Wolf Country*.

The authors are donating a portion of their royalty earnings from *The Ptarmigan's Dilemma* to the Nature Conservancy of Canada and to the World Wildlife Fund–Canada.